Design for Electrical and Computer Engineers
Theory, Concepts, and Practice

Design for Electrical and Computer Engineers
Theory, Concepts, and Practice

Ralph M. Ford
Penn State Erie, The Behrend College

Chris S. Coulston
Penn State Erie, The Behrend College

Boston   Burr Ridge, IL   Dubuque, IA   New York   San Francisco   St. Louis
Bangkok   Bogotá   Caracas   Kuala Lumpur   Lisbon   London   Madrid   Mexico City
Milan   Montreal   New Delhi   Santiago   Seoul   Singapore   Sydney   Taipei   Toronto

The McGraw·Hill Companies

McGraw-Hill Higher Education

DESIGN FOR ELECTRICAL AND COMPUTER ENGINEERS: THEORY, CONCEPTS, AND PRACTICE

Published by McGraw-Hill, a business unit of The McGraw-Hill Companies, Inc., 1221 Avenue of the Americas, New York, NY 10020. Copyright © 2008 by The McGraw-Hill Companies, Inc. All rights reserved. Previous edition © 2005. No part of this publication may be reproduced or distributed in any form or by any means, or stored in a database or retrieval system, without the prior written consent of The McGraw-Hill Companies, Inc., including, but not limited to, in any network or other electronic storage or transmission, or broadcast for distance learning.

Some ancillaries, including electronic and print components, may not be available to customers outside the United States.

This book is printed on acid-free paper.

3 4 5 6 7 8 9 0 DOC/DOC 0 9
ISBN 978–0–07–338035–3
MHID 0–07–338035–0

Global Publisher: *Raghothaman Srinivasan*
Executive Editor: *Michael Hackett*
Developmental Editor: *Darlene M. Schueller*
Director of Development: *Kristine Tibbetts*
Executive Marketing Manager: *Michael Weitz*
Senior Project Manager: *Kay J. Brimeyer*
Senior Production Supervisor: *Laura Fuller*
Executive Media Producer: *Linda Meehan Avenarius*
Designer: *John Joran*
Compositor: *Carlisle Publishing Services*
Typeface: 10/13 *Times Roman*
Printer: *R. R. Donnelley Crawfordsville, IN*

Library of Congress Cataloging-in-Publication Data

Ford, Ralph M. (Ralph Michael), 1965-
  Design for electrical and computer engineers : theory, concepts, and practice / Ralph M. Ford, Chris S. Coulston.
    p. cm.
  Includes index.
  ISBN 978-0-07-338035-3 --- ISBN 0-07-338035-0 (hard copy : alk. paper)
    1. Engineering design. 2. Electronic circuit design. 3. Electric engineering. I. Coulston, Chris S., 1967- II. Title.
TA174.F66 2008
621.3--dc22
                                   2007020263

www.mhhe.com

About the Authors

Ralph Ford obtained his Ph.D. and M.S. degrees in electrical engineering from the University of Arizona in 1994 and 1989 respectively. He obtained his B.S. in electrical engineering from Clarkson University in 1987. He worked for the IBM Microelectronics Division in East Fishkill, N.Y., from 1989–1991, where he developed machine vision systems to inspect electronic packaging modules for mainframe computers. Ralph also has experience working for IBM Data Systems and the Brookhaven National Laboratory. He joined the faculty at Penn State Erie, The Behrend College, in 1994. Ralph has experience teaching electronics and software design, as well as the capstone design course sequence in the electrical, computer, and software engineering programs. His research interests are in engineering design, image processing, machine vision, and signal processing. Ralph is currently Director of the School of Engineering at Penn State Behrend. He also serves as a program evaluator for ABET. He was a Fulbright Scholar at the Brno University of Technology in the Czech Republic in 2005.

Chris Coulston obtained his Ph.D. in computer science and engineering from Pennsylvania State University in 1999. He obtained his M.S. and B.S in computer engineering from Pennsylvania State University in 1994 and 1992 respectively. Chris has industry experience working for IBM in Manassas, Va., and Accu-Weather in State College, Pa. He joined the faculty at Penn State Erie, The Behrend College, in 1999. He has experience teaching design-oriented courses in digital systems, embedded systems, computer architecture, and database management systems. Chris' research interests are in Steiner tree routing algorithms and artificial life. He is currently an Associate Professor of Electrical and Computer Engineering at Penn State Behrend and also serves as chair of the program.

Table of Contents

Part I — The Engineering Design Process 1

Chapter 1 The Engineering Design Process 3
1.1 The Engineering Design Process 4
1.2 The World-Class Engineer 11
1.3 Book Overview 11
1.4 Summary and Further Reading 14
1.5 Problems 15

Chapter 2 Project Selection and Needs Identification 17
2.1 Engineering Design Projects 18
2.2 Sources of Project Ideas 19
2.3 Project Feasibility and Selection Criteria 20
2.4 Needs Identification 23
2.5 The Research Survey 27
2.6 Needs and Objectives Statements 30
2.7 Project Application: The Problem Statement 32
2.8 Summary and Further Reading 33
2.9 Problems 33

Chapter 3 The Requirements Specification 35
3.1 Overview of the Requirements Setting Process 36
3.2 Engineering Requirements 37
3.3 Developing the Requirements Specification 49
3.4 Requirements Case Studies 51
3.5 Advanced Requirements Analysis 57
3.6 Project Application: The Requirements Specification 61
3.7 Summary and Further Reading 62
3.8 Problems 62

Chapter 4 Concept Generation and Evaluation 65
4.1 Creativity 66
4.2 Concept Generation 71
4.3 Concept Evaluation 74
4.4 Project Application: Concept Generation and Evaluation 80
4.5 Summary and Further Reading 81
4.6 Problems 81

vii

Part II — Design Tools   85

Chapter 5	System Design I: Functional Decomposition	87
5.1	Bottom-Up and Top-Down Design	88
5.2	Functional Decomposition	89
5.3	Guidance	90
5.4	Application: Electronics Design	91
5.5	Application: Digital Design	95
5.6	Application: Software Design	98
5.7	Application: Thermometer Design	100
5.8	Coupling and Cohesion	105
5.9	Project Application: The Functional Design	107
5.10	Summary and Further Reading	107
5.11	Problems	108

Chapter 6	System Design II: Behavior Models	111
6.1	Models	112
6.2	State Diagrams	113
6.3	Flowcharts	115
6.4	Data Flow Diagrams	116
6.5	Entity Relationship Diagrams	119
6.6	The Unified Modeling Language	121
6.7	Project Application: Selecting Models	128
6.8	Summary and Further Reading	129
6.9	Problems	130

Chapter 7	Testing	135
7.1	Testing Principles	135
7.2	Constructing Tests	140
7.3	Case Study: Security Robot Design	146
7.4	Guidance	151
7.5	Summary and Further Reading	152
7.6	Problems	153

Chapter 8	System Reliability	155
8.1	Probability Theory Review	156
8.2	Reliability Prediction	161
8.3	System Reliability	172
8.4	Summary and Further Reading	177
8.5	Problems	177

Part III — Professional Skills   181

Chapter 9   Teams and Teamwork   183
9.1   What Is a Team?   184
9.2   Models of Team Development   185
9.3   Characteristics of Real Teams   187
9.4   Project Application: Team Process Guidelines   192
9.5   Summary and Further Reading   193
9.6   Problems   194

Chapter 10   Project Management   195
10.1   The Work Breakdown Structure   196
10.2   Network Diagrams   199
10.3   Gantt Charts   202
10.4   Cost Estimation   203
10.5   The Project Manager   207
10.6   Guidance   207
10.7   Project Application: The Project Plan   209
10.8   Summary and Further Reading   209
10.9   Problems   210

Chapter 11   Ethical and Legal Issues   213
11.1   Ethical Theory in a Nutshell   214
11.2   The IEEE Code of Ethics   216
11.3   Intellectual Property and Legal Issues   217
11.4   Handling Ethical Dilemmas   224
11.5   Case Study Analysis   226
11.6   Project Application: Incorporating Ethics in the Design Process   228
11.7   Summary and Further Reading   230
11.8   Problems   230

Chapter 12   Oral Presentations   235
12.1   How People Evaluate Presentations   236
12.2   Preparing the Presentation   237
12.3   Project Application: Design Presentations   240
12.4   Summary and Further Reading   243

Appendices

| Appendix A | Glossary | 245 |

| Appendix B | Decision Making with Analytical Hierarchy Process | 255 |

B.1 Applying AHP for Car Selection 256
B.2 Hierarchical Decision Criteria 260
B.3 Summary and Further Reading 262

| Appendix C | Component Failure Rate Data | 265 |

C.1 Environmental Use 265
C.2 Analog Components: Resistors and Capacitors 267
C.3 Microelectronic Devices 270

| Appendix D | Manufacturer Data Sheets | 281 |

1N4001 Rectifier Diode 282
2N3904 NPN Transistor 283
CD4001 Quad 2-Input NOR Gate 285
LM741 Operational Amplifier 288

| Appendix E | Design Project Case Study | 291 |

| References | 311 |

| Index | 317 |

Preface

This book is written for undergraduate students and teachers engaged in electrical and computer engineering (ECE) design projects, primarily in the senior year. The objective of the text is to provide a treatment of the design process in ECE with a sound academic basis that is integrated with practical application. This combination is necessary in design projects because students are expected to apply their theoretical knowledge to bring useful systems to reality. This topical integration is reflected in the subtitle of the book: *Theory, Concepts, and Practice*. Fundamental theories are developed whenever possible, such as in the chapters on functional design decomposition, system behavior, and design for reliability. Many aspects of the design process are based upon time-tested concepts that represent the generalization of successful practices and experience. These concepts are embodied in processes presented in the book, for example, in the chapters on needs identification and requirements development. Regardless of the topic, the goal is to apply the material to practical problems and design projects. Overall, we believe that this text is unique in providing a comprehensive design treatment for ECE, something that is sorely missing in the field. We hope that it will fill an important need as capstone design projects continue to grow in importance in engineering education.

We have found that there are three important pieces to completing a successful design project. The first is an understanding of the design process, the second is an understanding of how to apply technical design tools, and the third is successful application of professional skills. Design teams that effectively synthesize all three tend to be far more successful than those that don't. The book is organized into three parts that support each of these areas.

The first part of the book, *The Engineering Design Process*, embodies the steps required to take an idea from concept to successful design. At first, many students consider the design process to be obvious. Yet it is clear that failure to understand and follow a structured design process often leads to problems in development, if not outright failure. The design process is a theme that is woven throughout the text; however, its main emphasis is placed in the first four chapters. Chapter 1 is an introduction to design processes in different ECE application domains. Chapter 2 provides guidance on how to select projects and assess the needs of the customer or user. Depending upon how the design experience is structured, both students and faculty may be faced with the task of selecting the project concept. Further, one of the important issues in the engineering design is to understand that systems are developed for use by an end user, and if not designed to properly meet that need, they will likely fail. Chapter 3 explains how to develop the requirements specification along with methods for developing and documenting the requirements. Practical examples are provided to illustrate these methods. Chapter 4 presents concept generation and evaluation. A hallmark of design is that there are many potential solutions to the problem. Designers need to creatively explore the space of possible solutions and apply judgment to select the best one from the competing alternatives.

The second part of the book, *Design Tools*, presents important technical tools that ECE designers often draw upon. Chapter 5 emphasizes system engineering concepts including the well-known functional decomposition design technique and applications in a number of ECE problem domains. Chapter 6 provides methods for describing system behavior, such as flowcharts, state diagrams, and data flow diagrams and gives a brief overview of the Unified Modeling Language (UML). Chapter 7 covers important issues in testing and provides different viewpoints on testing throughout the development cycle. Chapter 8 addresses reliability theory in design, and reliability at both the component and system level is considered.

The third part of the book focuses on *Professional Skills*. Designing, building, and testing a system is a process that challenges the best teams, and requires good communication and project management skills. Chapter 9 provides guidance for effective teamwork. It provides an overview of pertinent research on teaming and distills it into a set of heuristics. Chapter 10 presents traditional elements of project planning, such as the work breakdown structure, network diagrams, and critical path estimation. It also addresses how to estimate worker needs for a design project. Chapter 11 addresses ethical considerations in both system design and professional practice. Case studies for ECE scenarios are examined and analyzed by using the IEEE (Institute of Electrical and Electronics Engineers) Code of Ethics as a basis. The book concludes with Chapter 12, which contains guidance for students preparing for oral presentations, often a part of capstone design projects.

Features of the Book

This book aims to guide students and faculty through the steps necessary for the successful execution of design projects. Some of the features are listed below.

- Each chapter provides a brief motivation for the material in the chapter followed by specific learning objectives.
- There are many examples throughout the book that demonstrate the application of the material.
- Each end-of-chapter problem has a different intention. Review problems demonstrate comprehension of the material in the chapter. Application problems require the solution of problems based upon the material learned in the chapter.

Design problems are directly applicable to design projects and are usually tied in with the Project Application section.
- Nearly all chapters contain a project application section that describes how to apply the material to a design project.
- Some chapters contain a guidance section that represents the author's advice on application of the material to a design project.
- Checklists are provided for helping students assess their work.
- There are many terms used in design whose meaning needs to be understood. The text contains a glossary with definitions of design terminology. The terms defined in the glossary (Appendix A) are indicated by ***bold-italic*** highlighting in the text.
- All chapters conclude with a summary and further reading section. The aim of the further reading portion is to provide pointers for those who want to delve deeper into the material presented.
- The book is structured to help programs demonstrate that they are meeting the ABET (accreditation board for engineering programs) accreditation criteria. It provides examples of how to address constraints and standards that must be considered in design projects. Furthermore, many of the professional skills topics, such as teamwork, ethics, and oral presentation ability, are directly related to the ABET Educational Outcomes. The requirements development methods presented in Chapter 3 are valuable tools for helping students perform on cross-functional teams where they must communicate with nonengineers.
- An instructor's manual is available at *www.mhhe.com/fordcoulston*. It contains not only solutions, but also guidance from the authors on teaching the material and managing student design teams. It is particularly important to provide advice to instructors since teaching design has unique challenges that are different from those of teaching engineering science–oriented courses that most faculty members are familiar with.
- PowerPoint™ presentations are available for instructors through the McGraw-Hill *www.mhhe.com/fordcoulston* website.
- There are a number of complete case study student projects available in electronic form for download by both students and instructors and available at *www.mhhe.com/fordcoulston*. These projects have been developed by using the processes provided in this book.

How to Use this Book

There are several common models for teaching capstone design, and this book has the flexibility to serve different needs. Particularly, chapters from Section III, Professional Skills, can be inserted as appropriate throughout the course. Recommended usage of the book for three different models of teaching a capstone design course is presented.

- **Model I.** This is a two-semester course sequence. In the first semester, students learn about design principles and start their capstone projects. This is the model that we follow. In the first semester the material in the book is covered in its entirety. The order of coverage is typically Chapters 1–3, 9, 4–6, 10–11, and 7–8. Chapter 9 (Teams and Teamwork) is covered immediately after the projects are identified and the teams are formed. Chapters 10 (Project Management) and 11 (Ethical and Legal Issues) are covered after the system design techniques in Chapters 5 and 6 are presented. Students are in a good position to create a project plan and address ethical issues in their designs after learning the more technical aspects of design. Chapter 12 (Oral Presentations) is assigned to students to read before their first oral presentation to the faculty. The course concludes with principles of testing and system reliability (Chapters 7 and 8). We assign a good number of end-of-chapter problems and have quizzes throughout the semester. By the end of the first semester, design teams are expected to have completed development of the requirements, the high-level or architectural design, and developed a project plan. In the second semester, student teams implement and test their designs under the guidance of a faculty advisor.
- **Model II.** This two-semester course sequence is similar to Model I with the difference being that the first semester is a lower credit course (often one credit) taught in a seminar format. In this model chapters can be selected to support the projects. Some of the core chapters for consideration are Chapters 1–5, which take the student from project selection to functional design, and Chapters 9–11 on teamwork, project management, and ethical issues. Other chapters could be covered at the instructor's discretion. The use of end-of-chapter problems would be limited, but the project application sections and example problems in the text would be useful in guiding students through their projects.
- **Model III.** This is a one-semester design sequence. Here, the book would be used to guide students through the design process. Chapters for consideration are 1–5 and 9–10, which provide the basics of design, teamwork, and project management. The project application sections and problems could be used as guidance for the project teams.

Acknowledgments

Undertaking this work has been a challenging experience and could not have been done without the support of many others. First, we thank our families for their support and patience. They have endured many hours and late evenings that we spent researching and writing. Melanie Ford is to be thanked for her diligent proofreading efforts. Bob

Simoneau, the former Director of the Penn State Behrend School of Engineering, has been a great supporter of the book and has also lent his time in reading and providing comments. Our school has a strong design culture, and this book would not have happened without that emphasis; our faculty colleagues need to be recognized for developing that culture. Jana Goodrich, Bill Lasher, and Rob Weissbach are three faculty members with whom we have collaborated on other courses and projects. They have influenced our thinking in this book, particularly in regard to project selection, requirements development, cost estimation, and teamwork. We must also recognize the great collaborative working environment that exists at Penn State Behrend, which has allowed this work to flourish. Our students have been patient in allowing us to experiment with different material in the class and on the projects. Examples of their work are included in the book and are greatly appreciated. John Wallberg contributed the disk drive diagnostics case study in Chapter 11, which we have found very useful for in-class discussions. John developed this while he was a student at MIT. Thanks to Anne Maloney for her copyediting of the first edition manuscript. The following individuals at McGraw-Hill have been very supportive and we thank them for their efforts to make this book a reality: Carlise Stembridge, Julie Kehrwald, Darlene Schueller, Craig Marty, Mike Hackett, Kay Brimeyer, Michael Weitz, Kris Tibbetts, Judi David, and John Joran.

Finally, we would like to thank the external reviewers of the book for their thorough reviews and valuable ideas. They are Ali Abul-Fadl (North Carolina A&T State University), Said Ahmed-Zaid (Boise State University), Frederick C. Berry (Rose-Hulman Institute of Technology), Mike Bright (Grove City College), Geoffrey Brooks (Florida State University Panama City Campus), Vikram Cariapa (Marquette University), Russell J. Clark (Saginaw Valley State University), Wils L. Cooley (West Virginia University), Bruce A. Ferguson (Rose-Hulman Institute of Technology), Ian Ferguson (Georgia Institute of Technology), D. J. Godfrey (U.S. Coast Guard Academy), Kemper Lewis (University of Buffalo–SUNY), David J. Nagel (The George Washington University), Michael Ruane (Boston University), Andrew Sterian (Grand Valley State University), and Vesna Zderic (The George Washington University).

We hope that you find this book valuable, and that it motivates you to create great designs. We welcome your comments and input. Please feel free to email us.

Ralph M. Ford, rmf7@psu.edu
Chris S. Coulston, csc104@psu.edu

Part I — The Engineering Design Process

Chapter 1 The Engineering Design Process

en-gi-neer (n) 1. *One versed in the design, construction, and use of machines.* 2. *One who employs the innovative and methodical application of scientific knowledge and technology to produce a device, system, or process, which is intended to satisfy human needs.* —American College Dictionary

Take a moment to read and analyze the key elements of the two definitions presented above. If you are an engineering student or practicing engineer, do you think that this definition applies to you? The first definition uses the terms *design* and *construction*. People like to think of themselves as designers. Why is that so? The answer may be in the combination of the term *construction*, and from the second definition, the idea of *innovation*. Applying innovation and creativity to produce something new is a wonderfully rewarding process. The great thing about being an engineer is that it allows you to be a creative designer. That is generally not the way the profession is viewed. What is the difference between engineering design and other types of design that are associated with creativity such as interior design, fashion design, or webpage design? The answer is supplied in the second definition which states "...methodical application of scientific knowledge and technology..." As an engineering student, you have studied a great deal of math, science, and fundamental technology, but probably have had limited exposure to creative and innovative design.

The definition also contains the somewhat contradictory terms *innovative* and *methodical*. If there is an established and methodical way of employing a scientific principle or process, it does not seem to allow much room for creativity and innovation. The truth is that the two concepts are in competition with each other, but a good engineer realizes this and utilizes both effectively. The definition also indicates that engineers design to satisfy human needs, an important, yet often overlooked point. That means that when designing systems, it is necessary to determine the user's needs and the ethical application of the technology.

This book aims to help electrical and computer engineers become effective designers, to better understand professional practices, and to provide guidance for executing design projects. This chapter presents the processes by which designs are realized, the characteristics of successful engineers, and an overview of the book.

Learning Objectives

By the end of this chapter, the reader should:
- Understand what is meant by engineering design.
- Understand the phases of the engineering design process.
- Be familiar with the attributes of successful engineers.
- Understand the objectives of this book.

1.1 The Engineering Design Process

ABET (formerly known as Accreditation Board of Engineering and Technology) provides the following definition of engineering design [ABE03].

> *Engineering design is the process of devising a system, component, or process to meet desired needs. It is a decision-making process (often iterative), in which the basic sciences, mathematics, and engineering sciences are applied to convert resources optimally to meet a stated objective. Among the fundamental elements of the design process are the establishment of objectives and criteria, synthesis, analysis, construction, testing, and evaluation.*

The definition indicates that, in engineering design, different phases of the process have to be revisited and the deliverables for each phase updated as necessary. Realistic problems are complex with many potential solutions; the goal is not to find just any solution, but the best one given the constraints and available resources. This requires the application of sound judgment, decision-making skills, and patience in constantly evaluating progress toward a solution. The definition identifies some common elements of the design process, such as establishment of criteria, synthesis, construction, and testing.

Design processes embody the steps required to take an idea from concept to realization of the final system, and are problem-solving methodologies that aim to develop a system that best meets the customer's need within given constraints. This is not all that different from some everyday processes, such as preparing dinner. Say you are hungry and need to eat dinner before you can go to see a movie that starts in one hour. The constraints are time, money, food, your tastes, and nutritional value if you are health-conscious. You brainstorm and come up with the options of making dinner at home, going to a restaurant, or buying something to eat at the theater. On the basis of these options, you then select the solution based on your evaluation of the best one. This is similar in philosophy to the stages of design processes where you have a problem to solve, constraints, and a number of potential solutions to select from.

A related term is known as the *product realization process*. The product realization process is broader in scope, including aspects such as entrepreneurship, market research, financial planning, product pricing, and market strategy. Many technologies have their own particular design processes that have evolved over time and have been found by practitioners in the field to be valuable. For example, different methodologies are applied in the design of integrated circuits (VLSI), embedded systems, and software systems, yet they all have some degree of commonality, such as requirements analysis, technical design, and system test. Design processes continue to evolve. One field in which this is particularly true is in software design because of the constantly changing nature of software and the special challenges that large software projects pose.

Cross [Cro00] identified two types of design processes—prescriptive and descriptive. As the name implies, *prescriptive design processes* set down an exact process, or systematic recipe, for realizing a system. Prescriptive design processes are often algorithmic in nature and expressed on flow charts with decision logic. An example of a prescriptive process is shown in Figure 1.1, which describes the front end of the design process where the problem and requirements are determined. A decision block is included where the requirements are examined to determine if they satisfy the needs of the problem. *Descriptive processes* are less formal, describing typical activities involved in realizing designs with less emphasis on exact sequencing. The distinction between descriptive and prescriptive processes is not always clear, however, and some processes may be considered more strongly associated with one property than the other. Cross makes an important point in stating that design processes are sometimes viewed as common sense and thus ignored, resulting in failed products. Cross cites two good reasons to adhere to design processes: (1) they formalize thought processes to ensure good practices are followed, leading to better and more innovative solutions, and (2) they keep all members of the team synchronized in terms of understanding where they are in the design process.

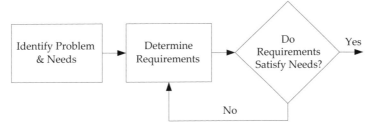

Figure 1.1 A prescriptive design process for problem identification and requirements selection.

A descriptive process that is widely applicable to design problems is shown in Figure 1.2. In a perfect world, the process starts with the identification of the problem, proceeds clockwise to research, followed the requirements phase, and so on until the system or device is delivered and goes into service (maintenance phase). This scenario is unrealistic, ignoring the iterative nature of design where the design team alternates between different phases as necessary. Consequently, links are inserted that allow transitions between all the different phases of the

6 Design for Electrical and Computer Engineers

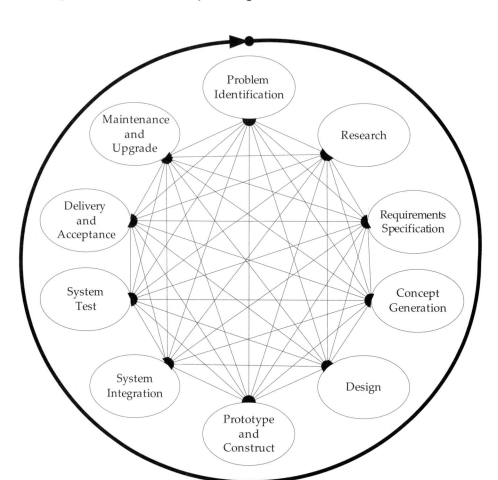

Figure 1.2 A descriptive overview of the design process.

design process. Of course, transitions between certain phases are unreasonable or very costly. It is virtually impossible to move directly from problem identification to system integration without developing a design concept first. It is much more likely for engineers to alternate between nearby phases in the process, such as problem identification, research, requirements specification, and concept generation. This does not mean that you can't move between phases that are not in close proximity in the model. For instance, the customer's needs may change while in the design phase, necessitating reevaluation of the needs, correction of the requirements specification, and system redesign—all at a substantial cost in time and money. Studies have shown that the cost required to correct errors or make changes increases exponentially as the project lifetime increases, as presented in Figure 1.3.

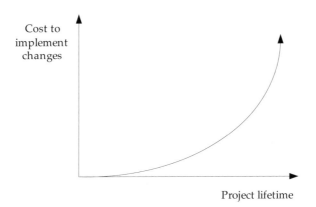

Figure 1.3 The cost to implement design changes increases exponentially with project lifetime.

1.1.1 Elements of the Design Process

Nearly all the phases of the design process in Figure 1.2 are covered in this book, with the exception of the maintenance phase. The objective of the first phase, *problem identification*, is to identify the problem and customer needs. This occurs in a variety of ways, from someone conceiving a new idea to a client coming to you with a problem to solve. In either case, it is important to determine the true needs for the product, device, or system (terms that are used interchangeably throughout the book and often referred to as systems). Failure to correctly identify the needs has negative ramifications for the entire process, typically resulting in costly redesigns, or even worse, abandonment of the project.

In the *research phase* the design team conducts research on the basic engineering and scientific principles, related technologies, and existing solutions. The objective is to become experts on the problem, save time and money by not reinventing the wheel, and be positioned to develop new and innovative solutions.

The *requirements specification* articulates what the system must do for it to be successful and to be accepted by the customer. It is important to focus on what the system must do, as opposed to how the solution will be implemented. This is challenging since engineers tend to focus on solutions and propose implementations early in the process. This is not surprising since engineering education focuses on solving problems instead of specifying them. The requirements are the mission statement that guides the entire project, and if properly developed, provide flexibility for creativity and innovation in developing solutions.

In *concept generation*, many possible solutions to the problem are developed. The hallmark of design is that it is open-ended, meaning that there are multiple solutions to the problem and the objective is to develop the one that best meets the requirements and satisfies the constraints. In this phase, wild creativity is encouraged, but it is ultimately tempered with critical evaluation of the competing alternatives.

In the *design phase*, the team iteratively develops a technical solution, ultimately producing a detailed system design. Upon its completion, all major systems and subsystems are identified and described using an appropriate model that depends upon the particular technology being employed.

In the *prototyping and construction phase*, different elements of the system are constructed and tested. In rapid prototyping, the objective is to model some aspect of the system, demonstrating functionality to be employed in the final realization. Many prototypes are discarded or modified as the system evolves—the idea is to experiment, demonstrate proof-of-concept principles, and improve understanding. Prototypes may be used anywhere in the process—you may present the client with prototypes after the concept generation phase, or they may be utilized in the design phase to test a design idea or as the final system is tested and developed.

During *system integration*, all of the subsystems are brought together to produce a complete working system. This phase is challenging and time-consuming since many different pieces of the design must be interfaced, and the team must work closely to make it all work. Care taken in the design phase to clearly communicate the functionality and interfaces between subsystems aids in system integration. System integration is closely tied to the *test phase*, where the overall system is tested to demonstrate that it meets the requirements.

Ultimately the system is *delivered* to the customer where it is likely that it will be tested by a mutually agreed upon process. Development does not necessarily end when the system goes into service, as it will likely enter the *maintenance phase* where it is maintained, upgraded to add new functionality, or design problems are corrected. Following and understanding the design process improves the probability of successful system development. The process is flexible, and the designer needs to transition between different phases in order to bring the system to realization. Design is an iterative process—you may not fully understand everything necessary in any given phase and have to revisit different steps as the system evolves. That is not a license for not trying to develop the best design you can on the first attempt—by all means do so—but realize that flexibility and a willingness to change the design are necessary.

1.1.2 Technology-Specific Design Processes

Different application domains have developed specialized processes for technology-specific design. One such example is VLSI (very large scale integration) design. A typical VLSI design process is shown in Figure 1.4 [Wol02]. In this model the system specification is used to develop the system architecture. The system architecture is composed of the major functional units that constitute an integrated circuit. Each functional unit is then designed at the gate logic level, which is subsequently designed at the circuit (transistor) level, and finally the circuit elements are laid out on the silicon chip. This is an excellent demonstration of the divide-and-conquer approach to design, where a complex system is broken down into lower levels of abstraction and each of these is further broken down until the design objectives are met.

Next, consider the design process for embedded computer systems shown in Figure 1.5. Embedded systems are combined hardware/software systems embedded into a larger system to perform dedicated application-specific operations. Embedded systems are employed in automobiles, DVD players, and digital cameras to name a few applications. Performance issues dominate embedded applications, and the designer needs to partition tasks between software and hardware to achieve optimum performance. This design process is somewhat prescriptive, with phases for requirements gathering, specifications, and architectural design. The process reflects the unique nature of embedded systems with separate software and hardware design blocks, married together by the interface design.

Figure 1.4 A process for integrated circuit (VLSI) design [Wol02].

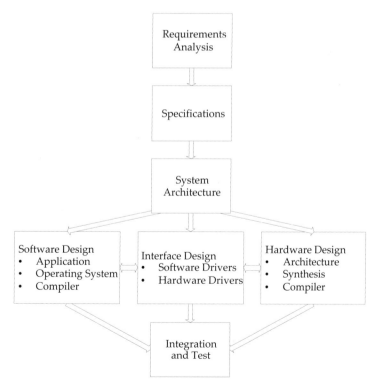

Figure 1.5 An embedded system design process [Ern97].

The field of software engineering is one in which the development of different design process models is still under considerable flux today. This is due to the complex nature of software and the failure of computer scientists and engineers to effectively develop high-quality software systems. There are many reasons why this is so. The sheer size of software programs may easily exceed one million lines of code written by many different software developers. One small mistake in those millions of lines of code can cause the system to fail. Another difficulty is in designing for upgrade and reuse of software. What if the needs change after the millions of lines of code are developed and one of the fundamental structures or objects needs to be upgraded?

The *waterfall model* shown in Figure 1.6 is one of the first proposed and most well-known software design processes. This is a prescriptive model since the development proceeds linearly from the first step where the user's needs are analyzed through the phases of specification development, design, test, and maintenance. This works for well-defined and moderately complex software applications, but fails as complexity grows because of the inability to move between phases. A more flexible and descriptive software design process is known as the *spiral model*, which is a cyclical process where phases are revisited as necessary [Som01]. *Extreme programming* is a more recent and controversial software development process, where relatively small teams of software developers rapidly develop software following some strict rules. Both the spiral model and extreme programming are examined in more detail in the end-of-chapter problems.

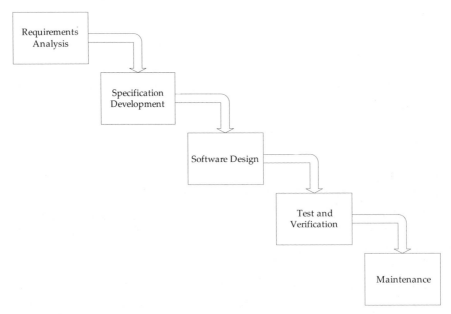

Figure 1.6 Waterfall software development process. In this model, development proceeds linearly from requirements analysis, through each subsequent phase, terminating with maintenance.

1.2 The World-Class Engineer

The ability to effectively design is important for engineers, requiring strong technical skills and an understanding of the design process. Yet, this ability in itself is not enough to become an effective practicing engineer. The Pennsylvania State University Leonhard Center for the Advancement of Engineering Education, in consultation with a number of industries, developed a description of what is referred to as a "World-Class Engineer" [Leo95]. Shown in Table 1.1, the description identifies the characteristics of successful engineers, and contains six major elements: (1) Aware of the World, (2) Solidly Grounded, (3) Technically Broad, (4) Effective in Group Operations, (5) Versatile, and (6) Customer Oriented. The description recognizes that engineers must be effective in group operations, since a majority of projects are carried out in teams. Not only that, many projects span multiple technical disciplines and are executed in multifunctional organizations that have diverse groups such as marketing, finance, human resources, technical support, and service. It also recognizes that an engineer must be versatile, innovative, understand ethical principles, and be customer oriented, important themes that are stressed throughout this book.

1.3 Book Overview

Consider the digital camera, the cellular phone, and the space shuttle, all complex systems that integrate a variety of technologies. A digital camera is the synthesis of an embedded electronics system, optics, a mechanical lens assembly, and the camera package itself. The embedded electronics contain an imaging sensor, a digital display, digital interface circuitry, flash memory storage, system control software, and the user interface. The challenges of integrating the components of such a system and having it record and transfer huge amounts of image data, within an acceptable time frame, are immense. Cellular phones are another good example of a complex system that represents a technology that has shrunk in size, but increased tremendously in functionality at the same time. They encompass digital data communications, an antenna, encryption for secure data transmission, a user interface display, and Internet connectivity. At the other end of the spectrum are large-scale space and military systems, such as the space shuttle. Despite the two shuttle accidents, the safety and reliability requirements of the space shuttle are incredibly high. Realizing such a system is accomplished by a tremendous number of people from many disciplines working for different organizations. All three of these technologies were developed by large teams that encompass multiple disciplines. The processes and practices employed in their development represent application of the fundamentals that this book hopes to cover. While you won't be building complete space shuttles by the end of this design course, you can expect to apply design principles that allow you to design and integrate a relatively complex system, maybe even a part of the space shuttle.

Table 1.1 The World-Class Engineer. (Copyright the Leonhard Center for the Advancement of Engineering Education, The Pennsylvania State University. Reprinted by permission.)

I. Aware of the World
- Sensitive to cultural differences, environmental concerns, and ethical principles.
- Alert to market opportunities (both high and low tech).
- Cognizant of competitive talents, work ethic, and motivation.

II. Solidly Grounded
- Thoroughly trained in the fundamentals of a selected engineering discipline.
- Has a historical perspective and remains aware of advances in science that can impact engineering.
- Realizes that knowledge doubles at breakneck speed and is prepared to continue learning throughout a career.

III. Technically Broad
- Understands that real-life problems are multidisciplinary.
- Thinks broadly, seeing an issue in a rich context of various alternatives, probabilities, etc., rather than as a narrow quest to find a single answer.
- Is conversant in several disciplines.
- Is trained in systems modeling and the identification of critical elements. Understands the need to design experiments to verify or extend analysis, as well as meet specification requirements.
- Is psychologically prepared to embrace any field necessary to solve the problem at hand.

IV. Effective in Group Operations
- Cooperative in an organization of individuals working toward a common creative goal that is often multidisciplinary and multifunctional in nature.
- Effective in written and oral communication.
- Willing to seek and use expert advice.
- Cognizant of the value of time and the need to make efficient use of the time in all phases of an endeavor.
- Understanding and respectful of the many facets of business operation—general management, marketing, finance, law, human resources, manufacturing, service, and especially quality.

V. Versatile
- Innovative in the development of products and services.
- Sees engineering as applicable to problem solving in general.
- Considers applying engineering beyond the typical employment focus of engineering graduates in the manufacturing industries, to the much broader economy (financial services, health care, transportation, etc.) where engineering skills could make a dramatic improvement in the productivity of those segments of the economy, which employ 80 percent of the U.S. population.

VI. Customer Oriented
- Realizes that finding and satisfying customers is the only guarantee of business success.
- Understands that products and services must excel in the test of cost-effectiveness in the global marketplace.

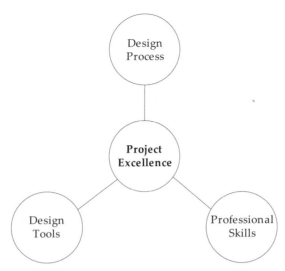

Figure 1.7 The guiding philosophy of this book. To achieve success in executing engineering and design projects, it takes an understanding of the design process, strong technical design tools, and professional skills.

The intention is to teach the application of design principles to computer and electrical engineers and to help prepare students for a professional career. The majority of engineering education is devoted to math, science, engineering science, and problem solving. They are important topics, required to enter this highly technical field. However, it is clear that there are other aspects beyond this that are equally important for success, including an understanding of system design, innovation, ethical principles, teamwork, and strong communication skills.

The book is divided into three parts: I–The Engineering Design Process, II–Design Tools, and III–Professional Skills. This is shown in Figure 1.7 as three separate, but related components that play a key role in achieving project excellence—the ability to complete a project, in an ethical manner that meets the customer's need, satisfies the constraints, and is clearly communicated to all involved. The chapters are decoupled as much as possible so that the reader can move between chapters as necessary. In Part I, the emphasis is on understanding and gaining experience in the different phases of the design process. The reader is guided through the steps of project identification, research, specification development, creative concept generation, and critical evaluation of competing solutions. Part II addresses topics that are often employed in design, including functional decomposition, description of system behavior, reliability, and testing. Part III addresses professional skills, including teamwork principles, project planning, ethics in design and the profession, and oral communication skills.

Here are a few thoughts to conclude the chapter and get started on the path to great designs. You are embarking on what will likely be a fun, challenging, sometimes frustrating, and ultimately rewarding journey. The systems that engineers work with continue to become increasingly complex and multidisciplinary in nature. The example problems presented in this book come from the fields of analog electronics, digital electronics, electrical systems theory, and software systems. These four areas constitute a significant problem-domain common to the education of most electrical and computer engineers. Finally, consider the quote below by Robert Hayes on the importance of design.

> *Fifteen years ago, companies competed on price. Today it's quality. Tomorrow it's design.—Robert H. Hayes, Harvard Business School, 1991.*

What is this saying? Well, it is clear that the world continues to move to a more knowledge-based society, where individuals and companies compete on the strength of their intellectual capital and ability to produce new and innovative products. That is what design is all about. It is not saying that price and quality are unimportant, they certainly are; in fact quality and reliability in design are part of this book. It is that quality and price are a given, and successful products will be distinguished by their design characteristics. The implication is that design will play a larger role in the development and success of products. The future Hayes predicted is now. Design is what distinguishes between products that are seen as commodities and those that are truly unique and profitable.

1.4 Summary and Further Reading

Engineering design is an iterative process in which the design team employs creativity and technical knowledge to develop a solution that best meets the end-users' needs within the constraints applied to the problem. There is no single design process that can be applied to all situations and technologies, but there are many common elements shared, regardless of the technology under consideration. In order to successfully bring designs to fruition, it takes a combination of design tools, professional skills, and a clear understanding of the process needed to complete designs. The objective of this book is to develop your proficiency in these areas so that you may become an effective engineer and achieve excellence in design projects.

Engineering Design Methods by Nigel Cross [Cro00] presents the differences between descriptive and prescriptive design processes, and covers a wide array of processes in more detail. It also discusses the cognitive characteristics of effective designers. There are many good books on software engineering process development methods. Software Engineering by Ian Sommerville [Som01] discusses the different software design process models, such as the waterfall and spiral models. This is also true of many modern software engineering texts. The original reference to the waterfall model is by Royce [Roy70]. The Art of Innovation by Michael Kelley [Kel01] describes the activities of well-known design company IDEO and is a highly readable description of their design practices. The ABC *Nightline* news program also produced an interesting segment on IDEO [ABC01] that can be purchased at the ABC website. The Circle of Innovation by Tom Peters [Pet97] is another popular book that provides his perspective on current trends in business and the importance of design.

1.5 Problems

1.1. In your own words, describe the difference between prescriptive and descriptive design processes. Cite examples of each.

1.2. Describe the relationship between the problem identification, research, and requirements specification phases of the design process.

1.3. Describe the relationship between the concept generation and design phases of the design process.

1.4. Construct a prescriptive design process for the problem identification, research, specification, concept, and design phases of the design process. The result should be a flowchart that contains decision blocks and iteration as necessary.

1.5. Describe the main differences between the VLSI and embedded system design processes.

1.6. Using the library or Internet, conduct research on the spiral software design process.

   a) Outline the significant elements of the spiral software design process.

   b) Describe the advantages and disadvantages of this relative to the waterfall model.

   Cite all references used.

1.7. Using the library or Internet, conduct research on the extreme programming design process.

   a) Outline the significant elements of the extreme programming paradigm.

   b) What are the pro and con arguments for this software development model?

   Cite all references used.

1.8. **Project Application.** In preparation for project and team selection, develop a personal inventory that includes a list of five favorite technologies or engineering subjects that you are interested in pursuing. Also, list the strengths and weaknesses that you bring to a project team.

Chapter 2 Project Selection and Needs Identification

For every problem there is a solution that is simple, neat, and wrong. —H. L. Mencken

Traditionally, companies have organized resources on the basis of functions such as accounting, engineering, finance, manufacturing, and marketing. It is often more effective to organize around projects that are of significant value and align resources to meet the needs of the project. This means that traditional departments and middle management are being de-emphasized and the role of projects is growing. Capstone design projects provide a great opportunity to gain experience in the management and execution of a project. One of the first and most important decisions encountered is selecting a project to pursue.

The objective of this chapter is to provide pragmatic guidance in the project selection phase. A description of design and engineering projects is presented, followed by advice on how projects can be selected by engineering students who wish to put design principles into practice. The chapter addresses how to identify the needs of the end user and provides guidance for conducting background research. All of this information is brought together in a problem statement that identifies the needs, the goals of the project, and research on the technology.

Learning Objectives

By the end of this chapter, the reader should:
- Have an understanding of the types of projects that electrical and computer engineers undertake.
- Understand and be able to apply criteria for project selection.
- Know how to determine, document, and rank end-user needs.
- Be aware of resources available for conducting research surveys.
- Have selected a project concept and developed a problem statement.

2.1 Engineering Design Projects

This section provides a classification of design and describes some of the types of projects undertaken by practicing engineers and those tackled in student projects. In reality, most projects don't fit neatly into the categories presented, but are some combination of them. The objective of a design project is to create a new *artifact* (system, component, or process) to meet a given need. Within the design domain there are different types of designs that are classified broadly into three categories of creative, variant, and routine designs [Cro00].

Creative designs represent new and innovative products. An example of a creative design is the Palm Pilot™ personal digital assistant (PDA). While the idea for the PDA had been around for awhile, earlier attempts at developing the technology, notably the Apple Newton, were unsuccessful. This was primarily a result of unreliable handwriting recognition that frustrated the user. However, Palm Computing had the creative idea to develop a simplified handwriting language, Graffiti, which eliminated the need for natural handwriting recognition. The Palm Pilot is a great example of a creative design—it is simple (four basic functions), fits in your pocket, and is easy to use. This innovation spawned a huge handheld computing industry.

Variant designs are variations of existing designs, where the intent is to improve performance or add features to an existing system. Many engineering projects fall into this category. For example, the objective may be to increase accuracy or system throughput.

Routine designs represent the design of devices for which theory and practice are well-developed. Examples are DC power supplies, analog and digital filters, and basic digital components such as adders and comparators. Routine designs are often components of more complex creative and variant designs.

Within these three categories of design, there are many different types of projects. *Systems engineering and systems integration projects* represent the synthesis of many subsystems into a larger system. They may be creative or variant designs, but have unique challenges since they are typically large and involve many people and technologies. Adherence to good design processes is important for their success. Engineers are often engaged in *systems test*, where the objective is to ensure that a system meets stated requirements and the needs of the user. Examples include the testing of systems for use in space and military environments.

The objective in *experimental design projects* is to design experimental procedures and apparatus for determining the characteristics of a system. For example, an engineering team may test a system under a variety of operating conditions. Example 2.3 presented later in the chapter is such a project, where the objective was to design a series of experiments to test the feasibility of gigabit Ethernet technology in a military environment. The test explored the impact of environmental factors such as temperature and vibration, and further used this data to estimate the operating lifetime of the Ethernet board. Upon completion of this project, the team made recommendations as to the allowable operating ranges of the technology.

The objective in *analysis projects* is to analyze some aspect of an existing system to improve or correct it. For example, a system or process may be failing in the field and the source of the failure unknown. Tools such as the failure mode effects and analysis technique may be applied in this situation to identify the sources of failure. In *technology evaluation projects*, technologies are assessed to determine if they can be used in a given application. This may be to determine if the technology can improve an existing system, or to characterize its operating performance.

The objective of a *research project* is to perform research or experiments with the goal of discovering or creating a new technology. The fundamental difference between this and other types of projects is that the ultimate outcomes are unknown. Most engineering research falls under the category of *applied research*. This refers to the creation of new technology or systems based on existing technology and theory developed from fundamental research. *Fundamental research* emphasizes the discovery of new scientific principles without necessarily having an intended application. Fundamental research is very valuable, but not typically a part of design projects.

2.2 Sources of Project Ideas

Depending upon your situation, you may have the opportunity to identify and select your project. The list below provides some places to search for project ideas:

- *Industry-sponsored projects.* Many companies will sponsor projects and are happy to do so, particularly if you have worked for them on an internship.
- *Engineers without Borders (www.ewb-usa.org).* This organization sponsors student projects to improve the quality of life in developing countries.
- *www.FreeRandD.com.* This is a clearinghouse for businesses and student teams to collaborate on projects. It allows businesses to post capstone project ideas for students to work on, while students can post resumes and project interests.
- *Your campus and local community.* In our school, a number of student teams have identified novel projects by asking other departments on campus for ideas. They have also been successful in approaching local community organizations for ideas, such as museums and research institutes.
- *Brainstorm.* Get together with a group of your peers and brainstorm on project ideas. You will be surprised at how many project ideas you can develop in a good brainstorming session (see Chapter 4). Do not only consider project ideas, but also brainstorm to identify problems that need solutions.

2.3 Project Feasibility and Selection Criteria

This section provides questions to consider when examining the feasibility of a project. George H. Heilmeier (an electrical engineer who has held positions as Chief Technology Officer of Texas Instruments, Director of the Defense Advanced Research Projects Agency, and CEO of Bellcore) developed a set of questions to answer when starting a new project [Sha94]. Heilmeier argued that all projects must be tied to the goals of the organization, and applied this by asking the following questions:

- What are you trying to do? Articulate your goals, using absolutely no jargon.
- How is it done today, and what are the limitations of current practice?
- What is new in your approach, and why do you think it will be successful?
- Who cares? If you are successful, what difference will it make?
- What are the risks and payoffs?
- How much will it cost? How long will it take?
- What are the midterm and final exams to check for success?

Heilmeier credits successful completion of projects that he managed to answering these questions up front and adhering to disciplined project management processes.

A second perspective is offered from an organizational project management viewpoint [Gra02] that provides the following criteria for project selection:

- *The project must be tied to the mission and vision of the organization.* Believe it or not, organizations often spend resources fruitlessly on projects that don't meet this criterion. To be fair, there is always risk associated with a project and it is sometimes hard to judge exactly how well a project meets this criterion. For engineers who are new to an organization, it is hard to judge a project's importance relative to the mission and goals, but if you find yourself in this situation, do not be afraid to ask some questions. Novices ask basic questions that are often overlooked by those who are highly experienced or intimately involved in a project.

- *Must have payback.* An economic analysis should be done to estimate if the project will make a profit. Much of this is outside the scope of this text, requiring marketing and financial analyses. Chapter 10 covers the basics of project cost estimation that will help in trying to answer this question.

- *Should have selection criteria.* Sound criteria for selecting among competing projects should be employed. Example 2.1 at the end of this section demonstrates the application of criteria in project selection.

- *Objectives of the Project should be SMART: specific, measurable, assignable, realistic, time-related.* Chapter 3 addresses how to determine project requirements that are specific and measurable. Assignable, realistic, and time-related all refer to project management aspects that are covered in Chapter 10. The objective is to develop tasks that are assigned to groups or individuals and realistically can be completed in the given timeframe.

The following example demonstrates how to apply a project selection model using a method known as the Analytical Hierarchy Process (AHP). AHP is a decision-making method that is described in Appendix B and is utilized frequently throughout the text—**the reader should read Appendix B prior to proceeding with this example**.

Example 2.1 A project selection model for capstone design.

Assume that you are part of a capstone design team that has the opportunity to select their project from competing project ideas. The steps in making a decision using AHP are to select the criteria that drive the decision, determine relative weights of the criteria, rate the alternatives (in this case project concepts) against the criteria to compute a weighted score for each of the alternatives, and then review the decision.

Step 1: Determine the selection criteria
To select the criteria, assume that the team brainstorms to determine the following criteria that interest the team members:

      A – Match to team skills
      B – Technical complexity
      C – Creativity
      D – Market potential
      E – Industry sponsorship

Step 2: Determine the criteria weightings
Assume the team applies the method of pairwise comparison to determine the weights as shown in Appendix B. In order to do so, the team systematically compares each criterion to all others, using the following scale of relative importance:

      1 = equal, 3 = moderate, 5 = strong, 7 = very strong, 9 = extreme.

Again, details of pairwise comparison are outlined in Appendix B. The results are below.

Criteria	A	B	C	D	E	Weight
A	1	5	5	3	3	0.52
B	1/5	1	3	1/3	1/3	0.12
C	1/5	1/3	1	1	3	0.09
D	1/3	3	1	1	5	0.18
E	1/3	3	1/3	1/5	1	0.09

This is an important step and one often overlooked—the team has identified what is important to it in project selection. It is clear that match to the team skills (criterion A) is most important, by a large margin, followed by market potential.

Step 3: Identify and rate alternatives relative to the criteria
Assume that the team identifies three potential projects ideas: 1—IEEE-sponsored robot competition, 2—industry-sponsored project to design a new test protocol, and 3—design of an item-finder device to help people locate lost items. Furthermore, the team goes through the process of rating each project relative to the criteria as outlined in Appendix B. These ratings are reflected in the decision matrix in the next step.

Step 4: Compute scores for the alternatives
The decision matrix below is constructed and the scores for the alternatives determined.

Selection Criteria	Weights	Alternatives		
		Project 1	Project 2	Project 3
A (Match to skills)	0.52	0.40	0.20	0.40
B (Technical Complexity)	0.12	0.40	0.30	0.30
C (Creativity)	0.09	0.45	0.20	0.35
D (Market potential)	0.18	0.05	0.35	0.60
E (Industry sponsorship)	0.09	0.00	1.0	0.00
Score		0.31	0.31	0.38

Step 5: Review the decision
Project 3 (item finder) is rated the highest among the three choices on the basis of the weights determined by the team members. It is a good match to the team skills, but also matches their desire to solve a problem with good market potential. The remaining two projects are rated about equal.

2.4 Needs Identification

Often a customer, client, or supervisor comes to you with a problem to solve and you must determine the needs or requirements for the solution to the problem. In other words, determine the *voice of the customer*. This seems like a simple statement—ask the customer what they want and you are done, right?

As an illustration, let's say a client comes to you with the following request: *The traffic at the front of campus is too congested. I would like you to design a new traffic lane for northbound traffic exiting at the intersection at the front of the college.* So you design this new lane and have it added to the intersection. However, you find out three months later that the traffic congestion has decreased a little bit, but it is still a significant problem. So what went wrong? Clearly you did what was asked of you, but the problem was not solved, meaning that you were solving the wrong problem. The real problem was to improve the flow of traffic at the entrance. In this case, the client gave both the problem and the solution all in one statement. That is fine if a careful feasibility study was done and it was known that the additional traffic lane would alleviate the problem, but that was not the case here. This hypothetical situation is not so far fetched and happens in practice via neglect to do the up-front research or because underlying assumptions change. The point is that the *correct* problem should be identified and solved.

It would be better if the client had simply asked to improve the traffic flow, providing the opportunity to analyze the situation and develop different design options. Some questions to be asked in this situation are: *How much additional traffic is there? At what times does this happen? Where is the traffic coming from? What is an acceptable waiting time at the intersection?* It may be that several new lanes are needed, or perhaps the sequencing of the traffic signals is wrong, or maybe a new entrance could be added for less cost and improved traffic flow.

The lesson is that customers often come with the problems and solution all wrapped up together. When this is done, the **design space**, the space of all possible solutions to the problem, is unnecessarily limited. Be ready to tactfully challenge the assumptions and ask questions to get to the root of the problem. Ask clarifying questions, analyze, pick apart the request, and focus on the problem, not the solution.

Researchers and practitioners have examined the problem of eliciting needs, and it is an important prerequisite for developing good engineering requirements specifications. Ulrich and Eppinger [Ulr03] proposed a process for obtaining the *voice of the customer* using the following five steps: (1) gather raw data from users; (2) interpret the raw data in terms of needs; (3) organize needs into a hierarchy; (4) determine the relative importance of the needs; and (5) review the outcomes and the process. Each of these steps is described in the following sections.

Step 1: Gather Raw Data from Users

DILBERT® by Scott Adams

Figure 2.1 The difficulties of communicating with the customer. (Dilbert © Scott Adams / Dist. by United Feature Syndicate, Inc.)

This is often accomplished via interviews with supervisors, key users, or people from the client organization (Figure 2.1). In cases where new products are being developed, focus groups are often employed. The advantage of interviews and focus groups is that they provide the opportunity for dialogue with the user where new ideas, concepts, and needs may emerge. Another option is direct observation, where the team goes out and examines the system in use and develops concepts for improving it. IDEO Corporation is an innovative and successful company that designs new products and systems. It relies heavily on direct observation as a technique for successfully developing innovative products [Kel01]. For example, IDEO was asked by a client to develop a new medical instrument for balloon angioplasty used in hospital operating rooms. A critical requirement from the user was that only one hand could be used to operate the device because the technician's other hand had to be free during the procedure. From direct observation, the IDEO design team found that even though the current system was designed for one-hand use, it was impractical, and the technicians actually used both hands. IDEO designed and developed a two-handed pump that not only worked better than the one handed pump, but was quieter, easier to read, and had increased precision. This is another example of the customer specifying the solution as part of the problem statement.

Ulrich and Eppinger provide the following questions to ask during an interview:

- When and why do you use this type of product (system)?
- Walk us through a typical session using the product.
- What do you like about the existing products?
- What do you dislike about the existing products?

- What issues do you consider when purchasing the product?
- What improvements would you make to the product?

Step 2: Interpret the Raw Data in Terms of Needs

In this step the raw data is translated into customer needs. The needs are expressed in terms of what the system must do (a requirement) as opposed to how it is done. Statements of the customer's needs are known as **marketing requirements** or **marketing specifications**. For example, "The system should have high-quality audio" is a need or marketing requirement from the customer regarding performance, but says nothing about how it will be achieved. Marketing requirements are short sentences that describe the need in the language of the customer. They typically do not have a numerical target and are described as a state of being for the system. Other examples of marketing requirements are, "The system should be easy to use," and "The system should be able survive a drop from the runner's height."

Step 3: Organize Needs into a Hierarchy

The marketing requirements are organized into a hierarchy of needs arranged from the most general to the most specific in successive levels of detail as required by the problem. It is organized by functional similarity, not as hierarchy of importance (that is the next step). This hierarchy is referred to as an **objective tree**. An example objective tree for a portable audio device intended for use by runners is shown in Figure 2.2. The three high-level objectives determined were high-quality audio, portable, and easy to use. Each of these is further subdivided into the characteristics that support the higher level. For example, portability is divided into the needs of lightweight, small, ergonomic, and the ability to operate in the environment. The environmental need is further expanded into needs that support it.

Step 4: Determine the Relative Importance of the Needs

The relative importance of the needs is based upon the user needs. As we saw in Example 2.1 and as presented in Appendix B, pairwise comparison is a good technique for determining relative importance and weighting of needs. In pairwise comparison, all needs are systematically compared to all other needs at the same level in the hierarchy. An example pairwise comparison table for this problem is shown in Table 2.1 with the resulting weights for each need indicated. This shows that portability is the most important need, followed by audio quality and ease-of-use. The weights are also reflected in the objective tree in Figure 2.2. In addition, the needs at each sublevel in the hierarchy are compared, the results of which are reflected in Figure 2.2. The rankings are used in later chapters to compare design alternatives.

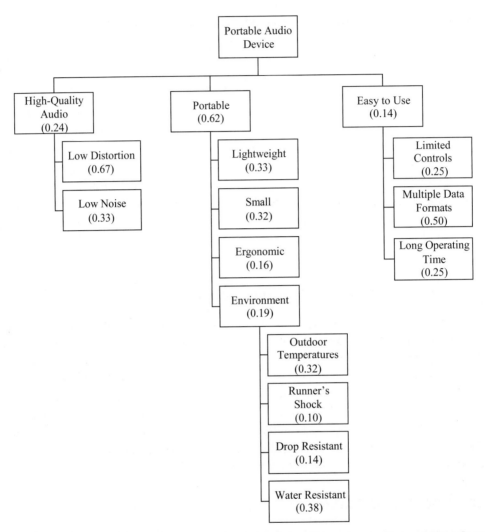

Figure 2.2 Objective tree for a portable audio device to be used by runners. The weights reflect the relative importance of needs at each level in the hierarchy as determined in Step 4 of the process.

Table 2.1 Pairwise comparison matrix for ranking the highest-level needs of the portable audio device. This comparison should be carried out for all levels of the objective tree.

	High-Quality Audio	Portable	Easy-to-Use	Weight
High-Quality Audio	1	1/3	2	0.24
Portable	3	1	4	0.62
Easy-to-Use	1/2	1/4	1	0.14

Step 5: Review the Outcomes and the Process

The design process and all of its subprocesses are methods for making good decisions, and this technique for needs identification is no different. There is a certain amount of subjectivity and judgment that goes into it; the end result should be reviewed to determine if it makes sense. The objective is to challenge assumptions, fully identify the problem, and make informed decisions.

The three outcomes of this process are the marketing requirements that identify the needs, an objective tree that provides a hierarchical representation of the needs, and a ranking of the relative importance of needs. This process may seem as though it does not apply to student design projects, but in reality it does. The questions in this chapter are certainly candidates to ask when working on company-sponsored projects. If it is not a company-sponsored project, the user needs should still be considered. For example, questions can be asked of friends and co-workers who are potential users of the system, focus groups can be formed, surveys administered, and Internet bulletin boards and discussion groups employed to gather this information.

2.5 The Research Survey

It is important to conduct a thorough research survey while defining the project concept. Failure to do so may translate into time and money spent reinventing the wheel, while not taking full advantage of existing components, knowledge, practices, and technology. During the research phase, competing systems and technologies are identified, and on the basis of this work, the project concept is refined, or in some cases abandoned. The character and strategy of the research survey is driven by the nature of the project. In general, the objective is to develop an understanding of the underlying scientific principles and demonstrate a familiarity with the state of the art in the particular field. Some questions to be answered in the research survey are:

- What is the basic theory behind the concept?
- How is it currently being done?

- What are the limitations of current designs or technology?
- What are the similarities and differences between your concept and existing technologies?
- Are there existing or patented technologies that may be relevant to the design? If so, what are they and why are they relevant?

Internet Searching

The Internet is a powerful, fast, and readily accessible source for conducting research. There are many excellent search engines for locating web resources, but understand that it is important to go beyond the well-known search engines and beyond the Internet in the survey.

One of the risks, and also one of the wonderful things, about the Internet is that virtually anybody can post information. It is important to analyze websites to ensure that they are reliable and credible. There are resources available that provide pointers on how to evaluate this credibility [Mci02, Sch98], and a little common sense goes a long way. One of the important things to look for is authorship—the author should be clearly identified and any affiliations listed. Carefully determine whether the information is subjective opinion or possibly a commercial for a product. Credible sites should provide references to original sources of material. Another step is to verify website content in print media or other reliable sources.

There are many search engines available, making the task of selecting one challenging. Also, there are different types of search engines: text (search for the text or keywords; subject heading or full-page text search), indexed (information categorized into directories), metasearch (engines that search other engines), and natural language processing (allowing natural language queries). Some search engines to try are *www.altavista.com*, *www.AskJeeves.com*, *www.google.com*, *www.kartoo.com*, and *www.yahoo.com*. The Librarian's Index to the Internet, *www.lii.org*, is a collection selected and evaluated by librarians, and according to their website, it is a "well-organized point of access for reliable, trustworthy, librarian-selected Internet resources." This information may change rapidly and represents current information at the time of publication.

Electrical and Computer Engineering Resources

Realize that the major search engines will not find all information available on the Internet. There are many websites with specialized search capabilities related to electrical and computer engineering design.

- EE Product Center, *www.EEProductCenter.com*. A website for locating electronic components and their manufacturers. It provides links to product datasheets and application notes. It has a keyword search engine and a tree structure search for finding components. For example, you can start with op amps and delve into subcategories such as precision and high-speed.
- Circuit Cellar, *www.CircuitCellar.com*. This companion website for the magazine is a great reference for designers. It emphasizes embedded systems and electronics projects with many tutorial articles and project ideas.
- Datasheet Catalog, *www.DatasheetCatalog.com*. A datasheet source for electronic components and semiconductors.
- Dr. Dobbs, *www.ddj.com*. The magazine and companion website are a resource for software developers that includes tips and tutorials.
- EE Times, *www.EETimes.com*. Industry newspaper for the electrical engineering field with information on current technology developments.
- Electronic Design Magazine, *www.EDNmag.com*. This is a free magazine for electrical design engineers that provides information on the latest products. The website has a number of categorized technical resources and a design ideas section.
- ON Semiconductor, *www.OnSemi.com*. ON Semiconductor is a supplier of semiconductors for a wide range of applications, with a particular emphasis on power management. The website has a searchable database of over 15,000 components, and provides guidelines for component selection based on different applications.
- The Thomas Register, *www.ThomasRegister.com*. This is a source for finding companies and products in North America. It allows searches for parts and equipment that may be used in a design project. It provides profiles of companies that meet the search criteria and describes the products they make.

In addition, most manufacturers of electronic components have websites providing product datasheets and application notes for their products. Application notes demonstrate how to use components in real applications. Examples are Dallas Semiconductor, Fairchild Semiconductor, Motorola, and Texas Instruments.

Government Resources

- U.S. Bureau of Labor Statistics, *http://stats.bls.gov*. This has valuable information on consumer spending, allowing one to determine things such as how much people spend and what they spend it on. It also profiles specific industries and forecasts employment in different industry sectors.

- U.S. Government Official WebPortal, *www.FirstGov.gov*. This is an entrance to all U.S. government web resources.
- U.S. Patent Office, *www.uspto.gov*. A searchable database of all patents back to 1790. Full text searches are available back to 1976 and full images back to 1790. It has information on the basics of patents, trademarks, and copyrights.

Journal and Conference Papers

The search should include journal and conference papers if technically detailed information on the latest theory or applications is needed.

- ACM (Association for Computing Machinery) Digital Library, *www.acm.org*. Provides abstracts (full text for subscribers) for ACM journals and conference proceedings.
- Compendex, *www.engineeringvillage2.org*. This provides indices to journal and conference papers in a broad scope of engineering fields, referencing material back to 1970.
- IEEE (Institute of Electrical and Electronics Engineers) Xplore Electronic Library, *www.ieee.org*. Provides abstracts (full text for subscribers) to all IEEE journals, transactions, magazines, and conference subscriptions published since 1988. Abstracts for all IEEE standards are publicly available.

2.6 Needs and Objectives Statements

Two parts of the problem statement are the needs and objectives statements. The *needs statement* identifies and motivates the need for the project and should:

- Briefly and clearly state the need being addressed.
- Not provide a solution to the problem.
- Provide supporting information collected as outlined in Section 2.4.
- Provide any supporting statistics and anecdotes that support the need.
- Describe current limitations.
- Describe supporting processes that are needed to understand the need. This is particularly important in industry-sponsored projects having specific needs that may not be clear to the average person.

The *objectives statement* typically ranges from one or two sentences to one or two paragraphs and should:

- Summarize what is being proposed to meet the need.

- Provide some preliminary design objectives (detailed requirements are developed later).
- Provide a preliminary description of the technical solution, avoiding a detailed description of the implementation. Often the input and output behavior of the system are described. The complete solution is not usually posted until after the engineering requirements are fully determined.

Example needs and objectives statements are provided in Examples 2.2–2.4.

Example 2.2 iPod Hands-Free Device Needs and Objectives. *Abstracted from the iPod Hands-Free Device Design Report by Al-Busaidi, Bellavia, and Roseborough [Alb07].*

Need: According to AppleInsider, approximately 10.3 million people owned iPods at the end of 2004 and many of the owners used them while operating their automobiles. The National Highway Traffic Safety Administration estimates that driver distraction is a contributing cause of 20 to 30 percent of all motor vehicle crashes—or 1.2 million accidents per year. One research study has estimated that driver inattention may cause as many as 10,000 deaths each year and approximately $40 billion in damages. iPods can present a distraction to drivers that is similar to that of cell phones in that the driver's attention is divided between controlling the steering wheel, watching the road, and navigating controls on the iPod. A system is needed to allow users to navigate among the music selections of their iPod without distracting their attention from the road.

Objective: The objective of this project is to design and prototype a device that will make the iPod safer to use while driving an automobile, by allowing hands-free control of the iPod. The device will interact with the user, using spoken English statements. The user will be able to issue simple voice commands to the device to control the operation of the iPod. In turn, the device will communicate information verbally, such as song titles that are displayed on the iPod screen, to the user.

Example 2.3 Experimental Design Problem Needs and Objectives. *Abstracted from the Intel Pro 1000XF Server Testing Design Report by Esek, Hunt, and Lewis. [Ese03].*

Need: Our industry sponsor is investigating the performance of commercial-grade gigabit Ethernet fiber optic equipment for computer data communications in a military environment. The proposed system will utilize an Intel Pro1000 XF server card. This is a harsh operating environment and its effects on the performance and lifetime of the equipment are unknown. The client wishes to understand how the military environment affects the optical power margin of the Intel Pro 1000 XF card and associated connectors and cabling.

Objective: The goal of this project is to design the experimental equipment and test procedures to determine the effects of temperature variations and vibration on the optical power margin and the operating lifespan of the system.

Example 2.4 Portable Aerial Surveillance Needs and Objectives. *Abstracted from the PASS Design Report by Andre, Kolb, and Thaler [And07].*

Need: Emergencies happen all across the world, all of the time. There are nearly 2,000,000 reported fires in the United States every year, and over 90 tactical activations of Pennsylvania's Special Emergency Response Team, which handles barricaded suspects and hostage situations. There have been over 100 documented riots in the United States in the past century, with the Los Angeles Riot alone causing $1 billion in damage. Having an aerial view of these situations would be a great benefit to the emergency workers on the ground. For example, police may have to monitor a large crowd or a hostage situation where aerial surveillance would allow them to observe the situation from a safe distance and use the footage as evidence in court. Firefighters could use aerial surveillance to examine fire-damaged buildings and search for victims through the windows of high-rise buildings. In large cities, emergency organizations often employ helicopters for aerial surveillance. However, in smaller rural towns, helicopters either take too long to reach the scene from a nearby city or they are too expensive to afford. The least expensive two-seat helicopters cost over $400,000, while new helicopters cost well over a million dollars with average operating costs of $400–$1000 per hour. There is a need for a low-cost aerial device that can provide emergency workers with overhead surveillance of emergency situations.

Objective: The objective of this project is to design a device that will provide emergency workers with a live aerial view of a situation at a cost that small municipalities can afford. The device will deploy rapidly and record and log video. The camera will also include pan and zoom functionality to make identification of victims and suspects easier.

2.7 Project Application: The Problem Statement

A format for a problem statement that integrates the elements of this chapter is as follows:
- *Need.* A statement that identifies the needs of the project.
- *Objective.* Describes the concept proposed to meet the needs identified.
- *Background.* A summary of the research survey on the relevant technologies and systems. The objective is to provide an introductory answer to the questions posed in Section 2.5. The length and content of this section varies, depending upon the project.

- *Marketing Requirements.* Short statements describe the user needs.
- *Objective Tree.* A hierarchical representation of the needs based on functional similarity with the relative weights of the needs identified.

2.8 Summary and Further Reading

This chapter addressed the types of projects that are often undertaken by engineers and provides guidance in terms of questions to ask when selecting a project. The success of design projects depends upon adequately determining the user's needs and desires for the system. A process developed by Ulrich and Eppinger for needs elicitation was presented. The three outcomes of this process are: (1) marketing requirements identifying the customer needs, (2) an objective tree that hierarchically represents the needs, and (3) a ranking of the relative importance of the needs. It is important to conduct research on the concept and related technologies, and pointers for conducting the research survey were provided. Finally, a format for a problem statement was presented that summarizes the needs, objectives, and research survey for a design project.

The works by Griffin and Hauser [Gri93] and Ulrich and Eppinger [Ulr03] are readable and more detailed discussions on how to obtain the voice of the customer. There are also other design books available that address how to identify needs and develop objective trees [Cro00]. Cagan and Vogel [Cag02] have proposed a process for product development known as iNPD (integrated new product development) and provide methods for navigating what they refer to as "the fuzzy front end of project definition."

2.9 Problems

2.1 In your own words, describe the differences between creative, variant, and routine designs.

2.2 List three guidelines that should be employed when selecting a project.

2.3 Assume a customer comes to you with the following request: *Design a mechanical arm to pick apples from a tree.* What are the assumptions in this statement? Rewrite the request to eliminate the assumptions. (This problem was originally posed by Edward DeBono [Deb70]).

2.4 Assume a customer comes to you with the following request: *Design an RS-232 networked personal computer measurement system to transmit voltage measurements from a remote location to a central server.* What are the assumptions of this statement? Develop a list of questions that you might ask the customer to further clarify the problem statement.

2.5 Describe what is meant by a marketing requirement.

2.6 What is the purpose of an objective tree and how is it developed?

2.7 The needs for a garage door opener have been determined to be: safety, speed, security, reliability, and low noise. Create a pairwise comparison matrix to determine the relative weights of the needs. Apply your judgment in making the relative comparisons.

2.8 Consider the design of an everyday consumer device such as computer printer, digital camera, electric screwdriver, or electric toothbrush. Determine the customer needs for the device selected. The deliverables should be: (1) marketing requirements, (2) an objective tree, and (3) a ranking of the customer needs using pairwise comparison.

2.9 **Project Application.** Select criteria to be applied for selecting a project concept as shown in Example 2.1 then brainstorm and search to generate project concepts. Rank the top three to five concepts against the criteria presented in Example 2.1.

2.10 **Project Application.** Determine the needs for the project selected. The result should be a list of marketing requirements, an objective tree, and a ranking of the needs.

2.11 **Project Application.** Conduct a research survey for your project using the guidance presented in Section 2.5. The result should be a report summarizing the results of the survey.

2.12 **Project Application.** Develop a problem statement for your project concept as outlined in Section 2.7. Apply the processes presented in the chapter as appropriate.

Chapter 3 The Requirements Specification

Specification (n) A detailed and exact statement of particulars, a statement fully describing something to be built. —American Heritage Dictionary

The requirements specification identifies those requirements that the design must satisfy in order for it to be successful. It is, in effect, the mission statement that drives all subsequent stages of development, and, when finished, should be a detailed and complete vision of the design goals. An effective requirements specification should identify all important requirements, yet provide enough flexibility for the design team to develop innovative solutions. It also serves as a communication tool for everyone involved in the design, such as engineering, marketing, and the client. All parties should agree to the requirements before further development proceeds. In some cases, the requirements specification serves as a legally binding agreement between the developers and the client.

A major challenge in developing the requirements is in many ways analogous to the proverbial "What came first, the chicken or the egg?" question. The final solution is analogous to the chicken and the requirements are analogous to the egg. In the beginning, the chicken is hidden inside the egg, yet the egg must be capable of describing what the chicken will become. The difficulty is that it is hard to develop the *what* for the requirement without already having solved the problem or created the chicken.

This chapter guides the reader through the process of developing a requirements specification and is organized as follows. First, the properties of an engineering requirement are defined and numerous examples for computer and electrical systems are provided. Then, the properties of the complete requirements specification (the collection of marketing and engineering requirements) are considered, followed by a number of case study examples. The chapter concludes with advanced methods of analyzing and refining requirements, utilizing tools such as the House of Quality.

Learning Objectives

By the end of this chapter, the reader should:
- Understand the properties of an engineering requirement and know how to develop well-formed requirements that meet the properties.
- Be familiar with engineering requirements that are commonly specified in electrical and computer systems.
- Understand the properties of the complete requirements specification, as well as know the steps to developing one.
- Be able to conduct advanced requirements analysis to identify design tradeoffs.

3.1 Overview of the Requirements Setting Process

The *requirements specification*, which is the focus of this chapter, is a collection of engineering and marketing requirements that a system must satisfy in order for it to meet the needs of the customer or end user. Figure 3.1 illustrates a process for developing a requirements specification that is from the IEEE Guide for Developing System Requirements Specifications [IEEE Std. 1233-1998]. This process is the focus of this chapter and there are three stakeholder groups in it—the customer, the environment, and the technical community. The input from the customer includes the marketing (raw) requirements that were addressed in Chapter 2. The environment introduces requirements in the form of constraints and standards that impact or limit the design. The input from the technical community is based upon the knowledge of engineers who are primarily responsible for design, implementation, testing, manufacturing, and maintenance of the system.

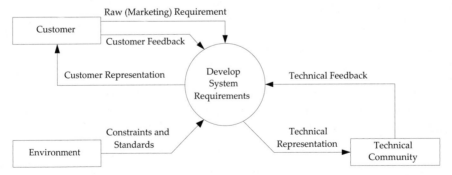

Figure 3.1 Requirements specification development processes from IEEE Std. 1233-1998. The three input sources to the process are the customer, environment, and technical community.

3.2 Engineering Requirements

Before developing the complete requirements specification, designers need to first determine individual engineering requirements. *Engineering requirements* are short statements that address a technical need of the design. A simple example is "The system should be able to supply 50 watts of power." This section identifies the desirable properties of engineering requirements, methods of identifying requirements, and provides numerous examples.

3.2.1 Properties of an Engineering Requirement

Each engineering requirement should meet the four properties below [IEEE Std. 1233-1998]:

1) *Abstract.* This means that a given requirement should specify *what* the system will do, not *how* it will be implemented. This is the chicken and egg problem described earlier. It is frequently the most difficult property to satisfy since designers often have a preconceived concept for the solution. Unless absolutely necessary, the requirements should say nothing about the implementation. For example, a requirement stating that a certain microcontroller (i.e., technology) *will* be used should be avoided. Admittedly, this is not always possible because of customer constraints or in cases where a system is being built upon preexisting technology. A common analogy used for the *what versus how* problem is that of designing a bridge. The requirement is to transport people from one side to other, without specifically stating the solution is a bridge, because another solution, like a ferry, may be a much more effective solution.

2) *Verifiable.* Verifiability means that there should be a way to measure or demonstrate that the requirement is met in the final system realization. Doing so allows the system to be tested or verified against the requirements. The idea is that if there is no way to verify that the requirement is met, then it should not be a requirement. Verifiability is used to answer the question of "Are we building the system correctly?"

3) *Unambiguous.* Each requirement should have a single unambiguous meaning and be stated with short complete sentences.

4) *Traceable.* Requirements should be traceable marketing requirements. If the design doesn't satisfy the customer's needs, it won't be successful.

Let's examine an example requirement for a robot whose objective is to navigate autonomously within a specified environment. Consider the following requirement:

> *The robot must have an average forward speed of 0.5 feet/second, a top speed of at least 1 foot/second, and the ability to accelerate from standstill to the average speed in under 1 second.*

Are the four properties for an engineering requirement met? In terms of the abstractness property, the answer is yes; it states what the system must do, not how it will be implemented. In terms of the second property, can the requirement be verified? Speed and acceleration are directly testable in the final realization, and thus it is verifiable. Is it unambiguous? It gives clear bounds for speed and acceleration. Finally, traceability can't be shown without the marketing requirements and is addressed later.

Now we analyze a second example requirement for the robot to see if it meets the properties:

> *The robot must employ IR sensors to sense its external environment and navigate autonomously with a battery life of 1 hour.*

This requirement is not abstract since it identifies part of the solution in terms of the sensor type and the fact that batteries must be used. It is somewhat ambiguous in that it should specify what is meant in terms of autonomous and the operating period. In terms of operating period, should it work for exactly 1 hour and stop, or is greater than an hour acceptable? Again, traceability can't be demonstrated without the marketing requirements. This requirement would be hard to verify without a good definition of what autonomous navigation in this context means. A better requirement would be:

> *The robot must navigate autonomously, with the aid of only landmarks in the specified environment, for a period of at least 1 hour.*

Realize that good requirements typically have two key elements in the statement—a description of *capability* and *condition*. Capability describes what the system must do and in the above requirement, that capability is autonomous navigation. Conditions are measurable or testable attributes of the capability and are critical for verification.

3.2.2 A Fifth Property—Realism

In addition to meeting the four properties, requirements should be realistic or justified. This is not defined in the IEEE standard as a property, but it is an important aspect that is often overlooked. To be realistic, there should be a way of demonstrating that the target is technically feasible. For example, a requirement could indicate that a robot should travel at a speed of 1,000,000 miles per hour, which could be verifiable, unambiguous, and abstract—yet, completely unachievable. Realistic targets can be determined with a little research, engineering know-how, creativity, or system modeling. One way to do this is to assume a solution for the final system—violating the abstractness property. For example, consider the design of a robot where some basic assumptions are made on the weight of the robot, the motors used, the wheel size, and the battery selected. An engineering model based upon these characteristics could be developed to predict performance and estimate realistic requirements. Alternatively, target requirements can be based upon an actual prototype, where a model or experimental system is developed to show that a particular requirement is feasible. This is how the technical community in Figure 3.1 feeds into the requirements process.

The use of benchmarking to identify similar systems and their performance provides a reference for realistic targets. It is generally hard to surpass the performance of well-developed products and systems on a first-generation design. An exception is with new and innovative approaches that allow you to surpass the competition. Competitive benchmarks may also be obtained from similar, but not necessarily identical, products. Experience working with a particular technology or previous generations of a system also provides guidance in selecting realistic targets. That being said, organizations wishing to gain or maintain a market edge often press the development team to achieve performance on new generations that were once believed to be unrealistic. Sometimes it just may not be feasible to determine the technical feasibility of requirements. In such cases, the requirements should have a certain amount of tolerance built into them and be updated as development proceeds.

3.2.3 Constraints

One of the inputs to the requirements process in Figure 3.1 is the environment, serving as the source of both constraints and standards. In reality, all engineering requirements impose some sort of constraint on a design, but in design a constraint is a special type of requirement. A *constraint* is a design decision imposed by the environment or a stakeholder that impacts or limits the design. Constraint requirements often violate the abstractness property. For example, a constraint requirement is

The system must use a PIC18F52™ microcontroller to implement processing functions.

This constraint requirement specifies how the system will be implemented. This could be because the project sponsor has developed a great deal of expertise using this particular microcontroller and does not want to spend the development time learning a new platform. Note that a number of other references define constraints to be synonymous with nonfunctional requirements (usually indicated as items that are not specifically functions). However, that terminology is avoided here since it is not well-defined nor universally accepted.

3.2.4 Standards

Standards are exactly what the name implies, a standard or established way of doing things that ensure interoperability. Without standards, the use of technology would be severely limited, if not downright impossible. Standards ensure that products work together, from home plumbing fixtures to the modules in a modern computer. Imagine if every computer manufacturer had its own communication standard, instead of following established protocols such as RS-232, TCP/IP, and USB—computers would have a hard time printing, sending email, instant messaging, or surfing the Internet! Furthermore, standards ensure the health and safety of products that people use every day. Identifying and following standards is an expected part of good engineering practice.

The focus in this chapter is on identifying standards that impact the requirements and ultimately the design. The question becomes, "What standards are relevant to your project and how do you use them?" There are different levels of interaction with standards that we denote as user, implementation, and development levels. At the *user level*, the standards are simply employed in the design, and detailed technical knowledge of the standard is typically not necessary. For example, when a component communicates to other devices, it is likely that a standard communication protocol is used. Other than configuring software or hardware to communicate with the standard, detailed knowledge of the standard isn't required. Another example would be in developing software to display digital images in a standardized format such as JPEG-2000 (Joint Photo Experts Group), in which case it is likely that existing software components would be used to read and display data in this format.

At the *implementation level*, details of the standard need to be understood. Standards at the implementation level are most likely to impact the design and the requirements. For example, when developing low-level drivers for computer peripherals, you need to become an expert on the underlying standard. Another example is reliability, where the requirement may be that "the system will have a reliability of 95% in 10 years." In this case a reliability standard, such as Military Handbook for Reliability Prediction of Equipment [MIL-HDBK 217F] may be employed, and its usage requires an understanding of both the reliability theory and the standard itself.

New standards are constantly being developed and existing ones modified, leading to the final level of interaction at the *development level*. Depending upon the standard, engineers from different organizations, professional societies, and corporations take part in the standards setting process. Many participants in this process are trying to gain a competitive advantage for their products and services.

It can be difficult to navigate the world of standards; they tend to be highly detailed and limited parts of a standard may apply to a project. In addition, many standards are costly to obtain, while some are freely distributed. The following is advice for identifying and employing standards. First, conduct research on applicable standards. Virtually all standards organizations maintain websites that provide basic information on their particular standards. The IEEE Xplore database is a good place to start, since it has a wide variety of standards and provides free searchable abstracts. Many companies and universities have subscriptions to databases of complete standards. Second, determine the expected level of interaction. From your analysis of the problem, do you foresee applying standards? Or will you need to develop an in-depth knowledge at the implementation level? In the latter case, you need to obtain detailed information on the applicable standards. Finally, you should consider asking your client. They may have its own internal standards and procedures to follow, and they may have experts on the applicable standards.

The list below identifies some of the types of standards that may be employed in a project and included in the requirements.

- *Safety*. Safety standards address how to design for safety and how to test products to ensure that they are safe.

- *Testing*. Testing standards are often related to safety, but are broader in scope. For example, standardized benchmark tests are used for comparing computational performance, one well-known standard being the SPEC (Standard Performance Evaluation Corporation) benchmarks.
- *Reliability*. Reliability standards address general reliability principles and design methods for different classes of systems. Another practical aspect is in the estimation of reliability of electronic systems, such as the IEEE and military reliability standards.
- *Communications*. They address how electronic systems communicate and transfer information, such as in computing, telephony, and satellite communications.
- *Data Formats*. Standard data formats ensure that systems and software can properly share information. Examples include image, video, and database standards.
- *Documentation*. There are standards for technical report documentation. In addition, there are standards for documenting processes and business practices, a well-known case being the ISO (International Standards Organization) 9000 and subsequent standards.
- *Design Methods*. Certain design techniques are standardized as well. Examples include software design methodologies, and the use of design languages such as the Hardware Description Language (HDL) and the Unified Modeling Language (UML).
- *Programming Languages*. Programming language syntax is standardized so that software maintains a level of portability between systems and compilers.
- *Connector Standards*. Standards for cable connections are common and should be followed to ensure that systems are easily interfaced and manufactured.
- *Metastandards*. Some standards are a combination of multiple standards known as metastandards. For example, the RS-232 standard is really a combination of a mechanical standard describing the connector physical dimensions connector, an electrical standard describing the voltages, a functional standard describing the pins and their function, and a procedural standard describing how entities communicate.

3.2.5 Identifying Engineering Requirements

There are many techniques for identifying requirements listed below [IEEE Std. 1233-1998]:
- Structured workshops and brainstorming sessions.
- Interviews, surveys, and questionnaires.
- Observation of processes or devices in use.
- Competitive benchmarking and market analysis.
- Prototyping and simulations.
- Research and technical documentation review.

Many requirements may be specified for a design, but knowing which to include is the challenge. The remainder of this section is a guide to describe the types of engineering requirements that may be specified for electrical and computer systems. Requirements in categories of performance and functionality are presented first as they are often critical, followed by an alphabetical grouping of a potpourri of other types requirements. **This taxonomy of requirements is by no means definitive or inclusive of all possibilities, and the design team needs to carefully determine those that are applicable to the particular situation. Careful attention must be given to the verifiability of requirements for the particular application.**

Performance

These requirements reflect a critical aspect of the performance of the system or device. They often are characterized by time, accuracy, throughput, or percentage error. The following is an example requirement that might be used in a security application with camera surveillance.

The system should detect 90% of all human faces in an image.

In order to verify this, a test might be constructed where the system is presented with a large database of face images that the system was not developed or trained with. The number of faces correctly detected would then be determined. Here is another example performance requirement for a system that measures part location:

The system should be able to measure part location to within ± 1 mm.

One way to verify this would be to take independent measurements of the system's ability to measure part location and compare them to the result of the system. The following is an example that could apply to software response time.

The system should retrieve the user data in no less than 3 seconds for 90% of requests and in a maximum of 6 seconds for all requests.

This could be verified by constructing a test where a large number of queries for user data are presented to the software under a variety of operating conditions and the response time measured. Yet another example is:

The system should be able to process video data at a rate of 30 frames per second.

This could be verified by providing an input video stream at the frame rate and testing to ensure that proper processing occurs. The test procedure would need to specify length and number of videos to test, issues that are addressed in Chapter 7. A final example of performance is one that could apply to electrical audio amplification.

The amplifier will have total harmonic distortion of less than 1%.

Total harmonic distortion is a measure that quantifies how closely an amplifier is able to replicate the original signal. This would likely be verified using laboratory instrumentation to measure the harmonic distortion in the output signal.

Functionality

These requirements describe the type of functions that a system should perform. Often, they provide inputs, outputs, and the transformation that the system will perform on the inputs. This is examined further in Chapter 5, which presents functional design techniques. The following is an example, where the input, ambient air temperature, is converted to a digital readout. It also has a performance aspect in that the accuracy is specified.

> *The system will convert ambient temperature to a digital readout of temperature with an accuracy of 1% over the measurement range.*

The following is an example from a real capstone project to develop a wireless mouse that is worn by the user and integrated into a glove.

> *The system will implement the left and right button functions of a standard mouse.*

The following are several functional examples for software systems.

> *The user shall be able to search all five company internal databases.*

> *The system will protect the user's identity with 128-bit encryption.*

Note that in these last two cases, verification would be by inspection.

Economic

Economic requirements include the costs associated with the development (design, production, maintenance) and sale of a system. They may also include the economic impact of the final system, such as how it will contribute to profits or save the user money. Two example economic requirements are below.

> *The costs for developing the system (labor and parts) should not exceed $50,000.*

> *The total parts and manufacturing costs cannot exceed $500 per unit.*

Energy

Virtually all systems consume and/or produce energy and thus have energy requirements. Energy consumption is the amount of power that a system consumes, and may be specified in terms of maximum, minimum, or average values. Example requirements are

> *The system will have an average power consumption of 500 mW.*

> *The system will have a peak current draw of 1 A.*

These requirements could be verified by measuring current and voltage draws under the different operating conditions, or by estimating the power drawn by all components in the system.

Operating lifetime addresses how long the system will operate from a given power source. For battery-powered devices, operating time is critical, and the lifetime for a given source may be an important requirement. An example of such a requirement is

> *The system will operate for a minimum of 3 hours without needing to be recharged.*

Source characteristics refer to the characteristics of the input and/or output sources, such as voltage, current, impedance, frequency, number of phases, and power requirements. An example requirement is

> *The system will operate from a 12 V source that supplies a maximum current of 300 mA.*

Environmental

These requirements address the impact of the design on the external environment and usage of the earth's resources. For example, energy usage is an important factor and example requirement is as follows:

> *The system will use 20% less energy than the industry average for similar products and qualify for U.S. Energy Star certification.*

Recyclability is the ability to dismantle a product into its constituent materials for reuse in other products. European countries have regulations on the recyclability of consumer products. In many cases, the producer of a product is responsible for its safe disposal once its service life is over. An example requirement is as follows:

> *50% of the modular components will be able to be repaired and reused in similar products.*

Health and Safety

The health and safety of anyone affected by the final product is an especially important consideration. For example, IEEE and ANSI standards provide guidance on safe levels for exposure to radio-frequency electric fields.

> *The system will not expose humans to unhealthy levels of electromagnetic radiation and will meet conditions for safe operation identified in ANSI Std. C95.1.*

There is a tendency to think that physical harm is not an issue in electrical and computer systems, but many electronic systems control mechanically moving parts. Consider the design of an automatic garage door system. An example constraint could be that

> *The door should stop moving if a person or object is detected in the door path.*

This could also translate into further engineering requirements on the amount of force on the door required to trigger it to stop. There are many safety standards, and two that are widely applied for consumer products are the UL (Underwriters Laboratory) and CE (Common European) standards. Examples are

> *The system will use only UL-approved components.*

> *The final system will be meet UL and CE standards and be tested at an independent laboratory for approval.*

Legal

Designs should not infringe upon existing patents, copyrights, and trademarks, particularly if the intention is to sell the product. Patent searches should be conducted, and search capabilities are available at the United States Patent Office website (*www.uspto.gov*). An example is

> *An intellectual property search will be conducted to ensure that there is no infringement on prior patents.*

This could be verified by having an external firm conduct the patent search and evaluate the design against existing intellectual property.

Security and privacy constraints apply to systems that handle sensitive data or personal communications. The ability of computing systems to withstand malicious attacks by hackers is another consideration, and the use of firewalls or other protective measures may be warranted. Examples are

> *The system will protect the user's identity with 128-bit encryption as required by law.*

Maintainability

The maintenance of the system being developed and compatibility with other systems are often considerations. Will the system be designed so that it can be reused in future applications? This is common in software development where the objective is to design modules that are reliable and flexible enough to be used in other applications. It is also a consideration in terms of the reusability of electronic or digital components in future system upgrades. An example reuse constraint is

> *The software should maintain downward capability and be able to use version 2 object libraries.*

After a product goes into service, it enters the maintenance phase, where it is maintained and upgraded. In software designs this is an important consideration, as software is regularly upgraded and maintained. On the hardware side, maintenance can be facilitated by the use of plug-in modules that are easily removed and replaced. Examples are

> *The system will initially be available to 100 users at five field locations, and within one year must be expanded to address usage by 5,000 users company-wide.*

> *The system should have a modular design such that failed components can be replaced by a technician in under 15 minutes.*

There may be internal restrictions on system development imposed by the company, based on their internal expertise and ability to maintain the system, such as the following constraint requirement.

The system will use only PIC microcontrollers.

Manufacturability

A prevalent product development paradigm that used to be employed in many engineering organizations was to "throw the design over the wall." What this meant is that the design and development team would create a new product and hand it off to the manufacturing team to produce (throw it over the wall and run), often without having considered the manufacturability of the product. The manufacturing team would then address how to produce the design, and in many cases could not do so without major redesign. Fortunately, this has given way to much better concurrent engineering practices where all aspects of product development are considered throughout the process. All of the examples presented here are constraints, in that they are external decisions that limit the design.

Size is a consideration in terms of the amount of space the final design will occupy, particularly if it has to be physically integrated with other components. An example constraint requirement is

The system must be manufactured on a circuit board with dimensions of no greater than 1" × 2".

The realization and portability of the system is a consideration. For example, with electronics, will it be built on a printed circuit board? Following design rules and file format guidelines is important for manufacturing printed circuit boards and integrated circuits. A chip foundry may require that integrated circuit layouts utilize certain file formats. Examples are below, which again are clear constraints on the design.

The system will be manufactured by using three-layer printed circuit board technology.

The product should be run on the Linux operating system.

The use of readily available parts, instead of low-volume or hard to find components, improves manufacturability, and an example is

The design shall incorporate only components that can be purchased through two of our main suppliers.

Operational

Operational requirements address the physical environment in which the system will operate. Characteristics could be temperature, humidity, electromagnetic radiation, shock, and vibration. Note that these can often be quite difficult to verify and may require specialized equipment to do so. An example temperature operational requirement is

> *The system should be able to operate in the temperature range of 0°C to 75°C.*

This could be tested via a test in an environmental chamber in which the operation of the system is tested over the complete temperature range. Alternatively, indirect verification is a possibility. For example, in the design, only components that are known to meet this operational requirement, as specified in their product datasheet, could be used.

Depending upon the customer needs, the system may be tested in an environmental chamber to verify that the requirement is met. Humidity is similar in concept to the temperature requirement, addressing the required ambient humidity range. A system may also need to be water resistant (withstand rain and snow) or waterproof (be submersible in water). An example requirement is

> *The system must be waterproof and operate while submersed in water.*

Be careful with the differences between waterproof (submersible in water) and water resistant, which indicates the ability to withstand outdoor elements such as rain and snow. For example, outdoor decorative lights are water-resistant, but not waterproof.

Depending upon the environment, the system may need to withstand vibrations. Bounds are typically specified in terms of frequency, magnitude, and duration of the vibration. An example requirement is

> *The system must be able to withstand vibrations of up to 60 Hz with a peak magnitude of 1 mm for a period of 1 minute.*

Electromagnetic interference *(EMI)* results from any electromagnetic energy that interferes or disturbs the operation of an electronic system. Electronic systems may produce electromagnetic radiation and limits may need to be placed upon the amount of radiation emitted. EMI is typically measured with specialized testing apparatus. Conversely, a system may need to be able to operate properly given a certain level of EMI.

The system may need to withstand a specified amount of shock and still operate. This may be measured in *g* force or via heuristics. An example requirement is

> *The system should withstand a drop from a height of 6 feet and still operate.*

Political

Political constraints address relationships to political, governmental, or union organizations. Examples include obtaining governmental approvals, resolving trade barriers, and determining the acceptance of systems for use in unionized environments. Examples are below.

> *The system will need to obtain FDA approval before it can be sold to medical users.*

> *The software will comply with the Digital Millennium Copyright Act.*

Reliability and Availability

This refers to the expected period of time that a system will operate properly. Measures of *reliability* include failure rates and mean time to failure. Estimation of system reliability is given detailed coverage in Chapter 8. The following is an example of a reliability requirement.

The system will have a reliability of 95% in 5 years.

This requirement means that 95% of the systems should be properly operating (have not failed) in 5 years. Direct measurement of reliability would not be possible, unless you are willing to wait 5 years to see how many systems fail, thus the use of estimation. This would require indirect verification using mathematical techniques to estimate the system reliability.

Availability is related to reliability, but addresses the amount of time that a system is available for operation. Example availability requirements are

The system will be operational 99% of the time.

The system will be operational from 4 AM to 10 PM, 365 days a year.

These might be hard to verify since it is determined for sure only when the system is deployed. Verification would have to address under which conditions the system would be tested to ensure this occurs.

Social and Cultural

This addresses aspects such as benefits, risks, and acceptance of products by the intended user or by society at large. For example, robots have tremendous benefits for improving product quality, while freeing people from dangerous and repetitive tasks. Yet when used in automation, they present the risk of displacing workers and causing job losses.

Many great products have fallen by the wayside because users were unwilling to accept it. An example is the early Apple Newton personal digital assistant, the first product of its kind. The fatal flaw was handwriting recognition; a training process was required for accurate recognition. This was not accepted by consumers, and Palm® Computing solved the problem by employing a simplified alphabet known as Graffiti. Graffiti was also seen as risky when it was being developed, but because of its simplicity it was accepted by consumers and the product became a huge success. In the mid-1980s, Phillips electronics released the laser disk player, which failed magnificently, but was far superior to VHS technology. Their failure was attributed to the cost of the players and disks compared to VHS. Fifteen years later DVDs, with the help of computers, reached a price point that now makes them preferable to VHS.

Will the system be used by engineers, technicians, laborers, doctors, lawyers, or the general public? Each group has its own culture, educational background, and willingness to accept innovations. Example requirements are

The product shall provide help menus to the user in either English or Spanish.

> *The software will be designed to easily be used by operators on the manufacturing floor. The software will be tested by a group of 25 operators and the average time to learn the basic functionality of the software will not exceed 8 hours.*

Usability

Usability requirements address the ease of use of a system. Although they are quite common, they are often difficult to verify. Usability can address how long it takes to learn the product and satisfaction by the end user or a group of users. To aid in verification conditions can be placed on the number of menus in the system, an estimated learning time, and number and types of errors the user is allowed to make. An example requirement is that

> *Users of the system should be able to learn 80% of its functionality within 2 hours.*

The method of verification would need to be clearly specified, such as, a group of 25 test users who have never used the product will be provided 2 hours to learn the product. Another example of a usability requirement is

> *The system will have a maximum of 20 functions and a maximum of two menus of depth.*

3.3 Developing the Requirements Specification

The requirements specification is the complete set of all system requirements. The steps in developing the requirements specification are to:

- Identify requirements from the customer, environment, and the technical community (focus of the previous section).
- Ensure the engineering requirements are well formed (meet the properties).
- Organize the requirements. Similar requirements should be presented together and relationships between engineering and marketing requirements identified. The collection of requirements should meet the properties identified in this section.
- Validate the requirements specification—which means all requirements are examined to ensure they meet the needs of the stakeholders.

3.3.1 Properties of the Requirements Specification

The desirable properties of the requirements specification are as follows [IEEE Std. 1233-1998]:

1) *Normalized (orthogonal) set.* There should be no overlap or redundancy between engineering requirements. A mathematical analogy is that of orthogonal vectors. For example, the x and y axes of the two-dimensional Cartesian space are orthogonal vectors, meaning that the projection (dot product) of one vector onto the other is zero. Ideally, all requirements should be orthogonal with no redundancy.

2) *Complete set.* A complete requirements specification addresses all of the needs of the end user and also those needs required for system implementation. Failure to define a complete set results in **underspecificity**, where not all needs are met.
3) *Consistent.* The engineering requirements should not be self-contradictory.
4) *Bounded.* The scope of the requirements specification should be identified. Determine the minimum acceptable bound for target values; going beyond what is necessary limits the design space of potential solutions. Applying unnecessary bounds results in **overspecificity**.
5) *Modifiable.* Requirements are typically considered to be evolutionary. This is because there are many unknowns at the start of a project, hence estimates for the requirements are made. The original requirements are known as **baseline requirements**. The estimates can change as development proceeds, as long as the changes are communicated to and agreed upon by all affected parties. Versions of the requirements should be tracked and identified as modifications take place.

3.3.2 Requirements Validation

An important property of an engineering requirement that we saw earlier was verifiability. Verifiability seeks to answer the question of whether or not the system is being developed correctly, or "Are we building the product correctly?" A related concept is that of validation, which seeks to answer the question "Are we building the correct product?" More formally, **validation** is the process of determining whether the system meets the needs of the user—is it valid? This is more general in scope than verification and more difficult to show. Requirements validation is usually carried out by reviews of the requirements by a team of people. Validation is demonstrated by being able to answer the following questions in the affirmative [Som01]:

- For each individual engineering requirement, are the traceability and verifiability properties met? Is each requirement realistic and technically feasible?
- For the requirements specification, are the properties of orthogonality, completeness, and consistency met?

A complete requirements specification includes all the requirements, both marketing and engineering, along with the relationships between them. The relationships between the engineering and marketing requirements need to be described to ensure that all the marketing requirements are being addressed by design. The relationship between the marketing and engineering requirements is called a mapping because, like a mathematical mapping, it defines which elements of the domain (marketing requirements) are associated with which elements of the range (engineering requirements).

3.4 Requirements Case Studies

This section presents case study examples of requirements specifications, most of which are from real capstone design projects. They are presented in a table format that presents each engineering requirement, the mapped marketing requirements (supporting the traceability property), and the justification for each requirement. The marketing requirements are summarized at the end of the table.

3.4.1 Case Study: Car Audio Amplifier

Table 3.1 presents the requirements specification for a car audio amplifier. This example was selected because of its simplicity and broad familiarity with this type of device. It will be expanded upon later.

Table 3.1 Requirements specification for an audio amplifier for use in an automobile.

Marketing Requirements	Engineering Requirements	Justification
1, 2, 4	1. The *total harmonic distortion* should be < 0.1%.	Based upon competitive benchmarking and existing amplifier technology. Class A, B, and AB amplifiers are able to obtain this level of THD.
1–4	2. Should be able to sustain an *output power* that averages ≥ 35 watts with a peak value of ≥ 70 watts.	This power range provides more than adequate sound throughout the automobile compartment. It is a sustainable output power for projected amplifier complexity.
2, 4	3. Should have an *efficiency* (η) > 40%.	Achievable with several different classes of power amplifiers.
3	4. *Average installation time* for the power and audio connections should not exceed 5 minutes.	Past trials using standard audio and power jacks demonstrate that this is a reasonable installation time.
1–4	5. The *dimensions* should not exceed 6″ × 8″ × 3″.	Fits under a typical car seat. Prior models and estimates show that all components should fit within this package size.
1–4	6. *Production cost* should not exceed $100.	This is based upon competitive market analysis and previous system designs. |//
Marketing Requirements 1. The system should have excellent sound quality. 2. The system should have high output power. 3. The system should be easy to install. 4. The system should have low cost.		

Audio power amplifiers are widely available devices, so the requirements were determined through competitive benchmarks and knowledge of amplifier circuit designs. The first engineering requirement directly impacts sound quality and is known as total harmonic distortion (THD). THD measures how closely the amplifier output signal follows the input signal. It is desirable for an amplifier to have a linear relation between input and output, where the output signal is identical to the input signal, except for an amplification factor. In reality, all amplifiers have some degree of nonlinearity or distortion. It is measured by applying a pure sinusoid as the input to the amplifier, which in the case of a perfectly linear amplifier produces a pure output sinusoid of the same frequency. Any nonlinearity introduces unwanted harmonic frequencies. THD represents the power of unwanted harmonics relative to the power of the fundamental sinusoid. THD is typically less than 1% for a good amplifier.

The second requirement, output power, is quantified in terms of both average and maximum values to minimize ambiguity. The third engineering requirement addresses the efficiency of the power transfer, or how much of the power consumed by the device is actually converted to audio power. The fourth engineering requirement, ease of installation, is perhaps easy to understand intuitively, and the expected installation time provides the condition for verification. The fifth requirement addresses the physical size of the device, and is important as it will need to be installed somewhere in the vehicle.

3.4.2 Case Study: iPod™ Hands-Free Device

Table 3.2 presents the requirements specification for a hands-free device whose intent is to allow a driver to communicate with an iPod™ audio player while driving. The problem statement was presented in Chapter 2 (Section 2.6).

Table 3.2 Requirements specification for the iPod™ hands-free device.

Marketing Requirements	Engineering Requirements	Justification
4, 6	1. System will *implement nine voice command* functions (menu, play/pause, previous, next, up, down, left, right, and select) and respond appropriately according to each command.	These are the basic nine commands that are used to control an iPod and will provide all functionality needed.
1, 3, 4, 7	2. The *time to respond* to voice commands and provide audio feedback should not exceed 3 seconds.	The system needs to provide convenient use by responding to the user inputs within a short time period. From research it was determined that the response time for the iPod is less than 1 second and an average voice

			recognition system requires 2 seconds to recognize commands.
4, 6	3.	The *accuracy* of the system in accepting voice commands will be between 95% and 98%.	Research demonstrates that this is a typical accuracy of voice recognition chips. Speaker-independent systems can achieve 95% and speaker-dependent up to 98%.
5, 6	4.	The system should be able to *operate* from a 12 V source and will draw a maximum of 150 mA.	The automobile provides 12 V DC. A current draw budget estimate was developed with potential components and 150 mA was an upper limit of current estimated.
5, 6, 7	5.	The *dimensions* of the prototype should not exceed 6″ × 4″ × 1.5″.	This system must be able to fit in a car compartment, somewhere between the seats. Estimate is based upon a size budget calculation using typical parts.

Marketing Requirements
1. Should not minimize or slow down the functional quality of the iPod.
2. User should be able to search for songs and artists and receive feedback on selection.
3. System should provide clear understandable speech.
4. System should be able to understand voice commands from user.
5. Should be able to fit and operate in an automobile.
6. Should be easy to use.
7. Should be portable.

To develop the marketing requirements, this team conducted an informal survey of students on campus, asking the target group what their desires for such a system would be. The first three engineering requirements are related to the important issues of the system functionality and performance. In order to develop justifications for some of the requirements, a prototype solution had to considered, although the solution had not been formally posed. For example, in order to estimate a time for responding to a user's command, some assumptions were made on the types of components that might be used in the design. The last two requirements are operational requirements, ensuring that the device will work in its intended environment.

3.4.3 Case Study: Gigabit Ethernet Card Testing

Table 3.3 presents an example requirements specification developed for the design of an experimental test setup [Ese03]. The problem statement for this example was presented in Chapter 2, where the objective was to design a system to test a gigabit Ethernet card for use in a harsh operating environment. In particular, the effects of temperature and vibration variations on the optical power margin, both of which impact the bit-error rate and the system performance, were determined.

Table 3.3 System requirements for a gigabit Ethernet card testing project.

Marketing Requirements	Engineering Requirements	Justification
1	Must be able to measure the *optical power output* with an *accuracy* of ±0.5 dB.	This is based upon commercially available optical power measurement instruments.
2	Must be able to measure the *optical power output* from 10°C to 55°C.	This range simulates the operating environment, and 55°C is the maximum operating temperature of the card.
2	The system must maintain *temperature accuracy* to within ± 1°C during all tests.	Based upon accuracy of commercially available test chambers.
3	Must be able to measure optical power over a *frequency range* from 4 Hz to 33 Hz in increments of 1 Hz.	The frequencies encountered in actual operation will not exceed this range.
3	The *peak vibration amplitude* should be 0.01 inches.	The amplitude in the operating environment will not exceed this value.
3	The card should be tested at a given frequency for a *duration* of 1 minute.	This exceeds the expected duration of vibration at given frequency that the system will encounter.
3	The vibration effects should be tested in *x, y,* and *z directions*.	The system will encounter vibrations in multiple directions. This will provide data on differences in directional variation due to vibration.
3	The experiment should determine *resonant frequency* to an accuracy of ±0.5 Hz.	This will provide data on worst-case vibration at the resonant frequency.

Marketing Requirements
1. The measurement of the optical power should be accurate.
2. It should measure the effects of temperature variations on optical power.
3. It should measure effects of vibration on the fiber optic connector and optical power output.

The marketing requirements were quite brief and direct, not surprising due to the fact the customer in this case was a group of engineers who had a good idea of what they wanted. The selection of engineering requirements was based upon characteristics of the operating environment, discussion with the engineers, and some educated guesswork. Let's consider some of the requirements, starting with the effects of temperature variation on the optical power output. The testing requirement on the temperature range was based on the operating environment, while the accuracy is driven by the test equipment. The requirements also address the vibration testing requirements, including the vibration frequency range, amplitude, and resonant frequency.

3.4.4 Case Study: Portable Aerial Surveillance System

The requirements specification for the portable aerial surveillance system (problem statement presented in Chapter 2) is shown in Table 3.4. This system is intended to provide police and emergency responders with a low-cost, easy-to-deploy aerial surveillance system.

Table 3.4 System requirements for the portable aerial surveillance system.

Marketing Requirements	Engineering Requirements	Justification
8	The device must not exceed a height of 150 feet above ground level for a moored device and 400 feet for a non-tethered remote controlled device.	To comply with FAA regulation 101.15.
8	The device must be lighted to ensure visibility at night and to maximize aircraft visibility during daytime.	This is required by FAA regulation 101.17.
8	The device will include a mechanism to force a controlled decent if it reaches the height ceiling for the device class.	This is required by the FAA regulation 101.19.
6	This device must remain airborne for a minimum of 4 hours.	Our contact suggested that the device should be deployable for an absolute minimum of two hours. Our comparison of commonly used police helicopter flight duration is 4 hours, so that is the flight time that we must match.
1	The device will cost less than $2,805.60/year to own and operate.	The Pennsylvania State Police Department owns and operates 7 helicopters at an average annual expense of $229,000/helicopter. The budget for the Pennsylvania State Police Department is $190 million dollars, making the helicopters 0.84% of the department's annual budget. The budget allotted to Erie county police departments was $16.7 million. 0.84% of Erie's annual budget would allow for $140,280 to be used for a helicopter. A device having less than 2% of the annual operating costs of a helicopter, and having many of the same surveillance capabilities would be considered a negligible expense.

3	The device will be able to be used in at least 14.1 m.p.h. winds and only move half of the width and half of the height of the current frame.	The device will be used outdoors and in non-ideal conditions in Erie, PA, where there is a 95% chance that winds will be at or below 14.1 mph and it must be able to keep an object or person in frame.
7	The device will allow visual recognition of both license plate text and state of origin from a minimum distance of 150 feet during daytime operation.	The recorded images and or video will be used as evidence during a trial. The device's maximum height is 150 feet, so the device will allow license plate recognition at an absolute minimum of that distance.
4, 5	The device must be operable by a single person. The flight and camera controls must be on a single unit with a joystick to control pitch and panning. Two slider potentiometers will be used to adjust the zoom and the focus.	A police officer dispatched alone in his or her cruiser should be able to launch and operate the device with no assistance.
2	The device must be smaller than the trunk of a police cruiser. The Chevrolet Impala has a trunk measuring 54 inches wide by 38 inches deep by 16 inches high.	The device will be transported in a police cruiser and must fit into the trunk of the vehicle.
4	The device will perform a power-on self test.	This device should ensure the operation of its electronic and/or wireless components before being put into the air.

Marketing Requirements
1. The product must have a low cost that a small municipality can afford.
2. The device must fit in an emergency vehicle, such as the trunk and be light enough to be removed by one officer.
3. The device must be durable.
4. The device must have a minimal learning curve.
5. The device must be operable by a single person.
6. The device must be able to remain in flight for at least four hours.
7. The device must have a video feed that can be recorded and usable in legal court.
8. The device must comply with FAA regulations.

This project was developed in conjunction with Penn State Behrend and Mercyhurst College's Institute for Intelligence Studies to meet a need that the institute identified. The student team met with a number of law enforcement officials to determine the needs. The first engineering requirement addresses the critical functionality that the system is to provide, while engineering requirements 2–5 address the performance and ability to work in the outdoor environment. Note the inclusion of a federal standard that drives the requirement on the maximum deployment height of the device. Requirement 7 addresses the cost, which can often be difficult to justify in student projects, but in this case the team has developed a clear justification.

3.5 Advanced Requirements Analysis

This section examines more advanced methods that are used to analyze and refine requirements. There are tradeoffs between the different requirements, and understanding them is valuable for refining the requirements themselves and developing solution concepts. This section addresses tradeoffs between engineering and marketing requirements, tradeoffs between engineering requirements themselves, and benchmarking. At the conclusion, all of this is integrated into the well-known House of Quality.

3.5.1 The Engineering-Marketing Tradeoff Matrix

This matrix identifies how engineering and marketing requirements impact each other. To demonstrate its construction, we continue to examine the automobile audio amplifier example from Table 3.1. The tradeoff matrix is shown in Table 3.5, where the marketing requirements constitute the row headings and the engineering requirements the column headings.

Table 3.5 Engineering-marketing tradeoff matrix for the audio amplifier (↑ = positive correlation, ↑↑ = strong positive correlation, ↓ = negative correlation, ↓↓ = strong negative correlation).

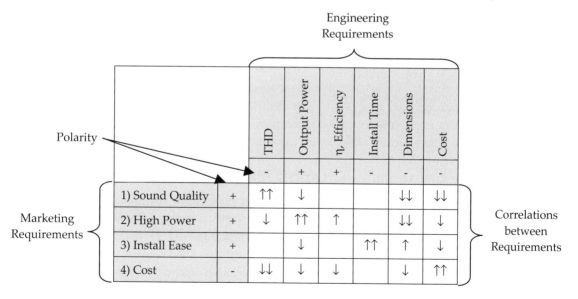

One of the first things to note is that each requirement has an associated polarity. A requirement with a positive/negative polarity, denoted with a +/− symbol, means that increasing/decreasing that requirement increases the desirability of the product, respectively.

A goal is considered a requirement with its polarity. For example, cost almost universally has a negative polarity because decreasing cost almost always makes a product more desirable.

The entries in the body of the matrix can be thought of as a correlation that measures the ability to achieve the marketing and engineering goals simultaneously. A positive correlation (↑) means that both goals can be simultaneously improved, while a negative correlation (↓) means that improving one will compromise the other. Not all correlations are of equal importance, the strength of the correlation being denoted by the number of arrows. Blank entries in the matrix mean that there is no correlation between the requirements.

To better understand the matrix, consider the entries in the top row associated with sound quality. The relationship between THD and sound quality is denoted by the double positive arrow. This relationship is interpreted as

> *The goal is to increase sound quality and decrease THD. There is a strong positive correlation between them, since decreasing THD increases sound quality.*

Another interesting case is the link between the goal of maximizing output power and the goal of maximizing sound quality (second entry in the top row). This entry is interpreted as

> *The goal is to increase sound quality and to increase output power, and there is negative correlation between them, since increasing output power will decrease sound quality.*

That is because electronics can be designed to achieve larger output power at the expense of sound quality. However, it gets a little more complicated, since output power can be increased without loss in sound quality, if more amplifier stages are employed. That increases the dimensions and cost, thus the relationships between sound quality, cost, and dimensions are identified in the first row. When creating the entries in the matrix, it should be assumed that only the associated requirements can vary and that all others are held constant. When finished, the matrix allows a quick and easy reading of the tradeoffs between engineering and marketing requirements. This example demonstrates the complex nature of a seemingly simple device, and provides a much clearer picture of the design tradeoffs involved.

3.5.2 The Engineering Tradeoff Matrix

The complex interactions regarding output power in the previous section illustrates the need to examine the tradeoffs between the engineering requirements, which are shown in Table 3.6. In this table the engineering requirements constitute the headings for both the row and column entries. Only the entries above the upper diagonal elements are filled in due to the redundancy of the lower diagonal elements. Again, positive and negative correlations are indicated along with the strength of correlation.

Let's examine the tradeoffs involved with output power. High output power can be achieved at the expense of THD as shown in the first row of the table. The second row indicates that there is a positive correlation between efficiency and output power, since the more efficient an amplifier is, the more power it can deliver. There is a negative correlation between the dimensions and power, since larger parts and greater surface area aid in dissipating more power. Finally, there is a negative correlation between the output power and cost because of the greater size and number of parts needed to achieve higher power.

Table 3.6 The engineering tradeoff matrix for the audio amplifier (↑ = positive correlation, ↓ = negative correlation).

		THD	Output Power	η, Efficiency	Install Time	Dimensions	Cost
		−	+	+	−	−	−
THD	−		↓			↓	↓
Output Power	+			↑		↓	↓
η, Efficiency	+					↑	↓
Install Time	−					↓	
Dimensions	−						↓
Cost	−						

3.5.3 Competitive Benchmarks

Competitive benchmarking helps to select targets for the engineering requirements. By analyzing competing systems, a better understanding is gained of what is realistic and where the design may potentially outperform the competition. The benchmark table lists the requirements in the row headings and the competitors in the column headings as shown in Table 3.7.

Table 3.7 Competitive benchmarks for the audio amplifier.

	Apex Audio	Monster Amps	Our Design
THD	0.05%	0.15%	0.1%
Power	30 W	50 W	35 W
Efficiency	70%	30%	40%
Cost	$250	$120	$100

3.5.4 The House of Quality

A well-known tool for developing requirements is the House of Quality (HOQ). The HOQ is part of a product development process known as quality functional deployment (QFD) that is widely used in industry. QFD is a series of processes for product development that incorporate the needs of the customer throughout the system life cycle. It encompasses design, manufacturing, sales, and marketing. QFD is characterized by a series of matrices that have a visual appearance similar to that of a house. The matrices relate different aspects of the development process and are effective for communicating between different units in an organization. There are houses for different phases of product development, but here the focus is on using the HOQ for the requirements specification. A HOQ for the audio amplifier example is shown in Figure 3.2.

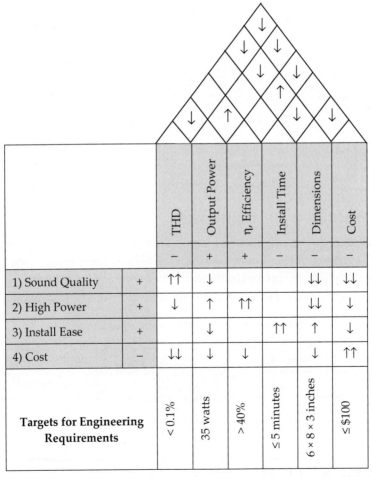

Figure 3.2 The complete House of Quality for the audio amplifier example. This integrates the information in Tables 3.5, 3.6, and 3.7.

It contains all of the elements that we have addressed so far—marketing requirements, engineering requirements, engineering-marketing tradeoffs, engineering tradeoffs, and the target values for the engineering requirements. The HOQ is presented for completeness, but is redundant since it contains all of the information already presented in Tables 3.5–3.7. The HOQ also becomes visually overwhelming and hard to read as problem complexity grows.

3.6 Project Application: The Requirements Specification

The following is a recommended format for a requirements document that integrates the problem statement from Chapter 2.

- *Needs, Objectives, and Background*. Include the elements from the problem statement in Chapter 2.
- *Requirements*. Identify the marketing requirements, engineering requirements, and justification in a table format (see Tables 3.1–3.4). Supplement this with tradeoff matrices and competitive benchmarks as necessary.

Table 3.8 presents a self-assessment checklist for the requirements specification.

Table 3.8 Self-assessment checklist for the requirements specification (1 = strongly disagree, 2 = disagree, 3 = neutral, 4 = agree, 5 = strongly agree).

Engineering Requirements	Score
Each engineering requirement is abstract.	
Each engineering requirement is verifiable.	
Each engineering requirement is unambiguous and written as a concise statement.	
Each engineering requirement can be traced to a user need.	
Each engineering requirement is realistic and has a justification provided.	
Standards and constraints applicable to the project have been identified and included.	
The Requirements Specification	
The requirements are normalized, with minimal redundancy and overlap.	
The engineering requirements are organized by similarity.	
The requirements are complete, addressing all needs.	
The requirements are bounded (not overspecified).	
The requirements have been validated and agreed upon by all stakeholders.	

3.7 Summary and Further Reading

This chapter presented a process for developing the requirements specification, which consists of identifying the requirements from the user, environment, and input from the technical community. The desirable properties of engineering requirements and the complete requirements specification were presented. The verification of a requirement is particularly important, as it helps in answering if the system is being built correctly. Requirements validation addresses whether the requirements meet the needs of the user, or if the correct product is being designed. Tools for benchmarking and analyzing the tradeoffs between requirements were given. Proper determination of the requirements significantly influences all subsequent phases of the design, thus the final requirements document should be agreed upon by all stakeholders.

The processes presented here were developed from research in the field and the authors' teaching experiences. Pugh [Pug90] presents a good perspective on identifying requirements and constraints, although with more emphasis on mechanical systems. The article by Robert Abler [Abl91] is a short primer that provides good advice on how to develop specifications and it overlaps with the properties presented in the IEEE Standard 1233 [IEEE Std. 1233-1998].

The HOQ technique was originally developed by Hauser and Clausing [Hau88] and has gained wide acceptance. Their original article provides a case study of the technique applied to the design of automobile door seals as implemented by Toyota Motor Corporation. Ulrich and Eppinger [Ulr03] present a good perspective on developing specifications employing the QFD techniques and the HOQ with an emphasis on the voice of the customer.

3.8 Problems

3.1 Briefly describe the four properties of an engineering requirement.

3.2 Identify the three levels of standards usage and what is meant by each one.

3.3 For each of the engineering requirements below, determine if it meets the properties of abstractness, verifiable, unambiguous, and realistic. If a requirement does not satisfy the properties, restate it so that it does:
   a) The TV remote control will be easy to use.
   b) The robot will identify objects in its path using ultrasonic sensors.
   c) The car audio amplifier will be encased in aluminum and will operate in the automobile environment.
   d) The audio amplifier will have a total harmonic distortion that is less than 2%.
   e) The robot will be able to move at a speed of 1 foot/second in any direction.
   f) The system will employ smart power monitoring technology to achieve ultralow power consumption.

g) The system shall be easy to use by a 12 year old.
h) The robot must remain operational for 50 years.

3.4 Provide three example engineering requirements that are technically verifiable, but not realistic.

3.5 Describe the difference between *verification* and *validation*.

3.6 Explain how *validation* is performed for a requirements specification.

3.7 Provide an example of a project (real or fictitious) where verification is successful, but validation is unsuccessful.

3.8 Consider the design of a common device such as an audio CD player, an electric toothbrush, or a laptop computer (or another device that you select). Identify potential marketing and engineering requirements. Consider those categories presented in Section 3.2, as well as any others that are applicable to the problem. You do not need to select the target values, but should identify the measures and units. Present the requirements in a table format as in Table 3.1.

3.9 Develop a marketing-engineering tradeoff matrix for the device selected in Problem 3.8.

3.10 Develop an engineering tradeoff matrix for the device selected in Problem 3.8.

3.11 Develop a list of potential standards that would apply to one of the devices proposed in Problem 3.8, and for each indicate how it would apply to the design.

3.12 **Project Application.** Develop a complete requirements document for your project as outlined in Section 3.6. Make sure that the engineering requirements meet the five properties identified in the chapter. The team should complete the self-assessment checklist in Table 3.8.

Chapter 4 Concept Generation and Evaluation

Creativity is a great motivator because it makes people interested in what they are doing. Creativity gives hope that there can be a worthwhile idea. Creativity gives the possibility of some sort of achievement to everyone. Creativity makes life more fun and more interesting.— Edward DeBono

In developing a design, it is important to explore many potential solutions and select the best one from them. Too often a single concept is generated and is the only one pursued, the unfortunate result being that potentially better solutions are not considered. When confronted with a problem, engineers must explore different concepts, critically evaluate them, and be able to defend the decisions that led to a particular solution. Two key thought processes employed are creativity and judgment. Creativity involves the generation of novel concepts, while judgment is applied to evaluate and select the best solution for the problem. Creativity and judgment appear to be inherent individual qualities that can't be taught. That is to some extent true, but with practice and application of formal techniques, they can be improved.

It is important to distinguish between innovation and creativity. Creativity refers to the ability to develop new ideas, while innovation is the ability to bring creative ideas to reality. Innovation is valued by companies since new products and services are often their lifeblood. That is why many make it a priority to hire engineers who can bring creativity to the design process. This chapter addresses creativity, concept generation, and evaluation in design. The first part describes barriers to creative thought, followed by strategies for overcoming them and enhancing creativity. Next, methods for concept generation are presented, followed by techniques for concept evaluation.

Learning Objectives

By the end of this chapter, the reader should:

- Understand the importance of creativity, innovation, concept generation, and concept evaluation in engineering design.
- Be familiar with the barriers that hinder creativity.
- Be able to apply strategies and formal methods for concept generation.
- Be able to apply techniques for the evaluation of design concepts.

4.1 Creativity

Is creativity something that is inherent in the individual or something that can be learned? It appears that both are true; some individuals are naturally more creative than others, yet people can enhance their creativity with conscious effort and practice. This section examines barriers to creativity, different thinking modes, and strategies for enhancing creativity. One of the ways to spark creativity is to solve puzzles. To get into the creative spirit the reader can try to solve the puzzles presented in Figure 4.1.

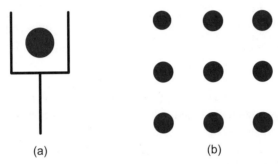

Figure 4.1 (a) The shovel problem. Think of this as a shovel with a coin on the spade. The objective is to move two lines so that the coin is no longer in the spade, but there is still a shovel. (b) The nine dot problem. Draw four connected straight lines that pass through all nine dots.

4.1.1 Barriers to Creativity

James L. Adams, an engineer and former professor at Stanford University, has researched innovation in technical domains. He examined the barriers to creativity and classified them into the following four types: (1) perceptual blocks, (2) emotional blocks, (3) cultural and environmental blocks, and (4) intellectual and expressive blocks [Ada01].

Perceptual blocks are those that prevent people from clearly seeing the problem for what it is. A common perceptual block is the tendency to delimit the problem space, or in other words, to put constraints on the problem that don't exist. Have you solved the puzzles shown in Figure 4.1 yet? If not, it is possible that you are placing constraints on the problems that don't exist. Knowing that this is the case, you may want to go back and try again. Another example of a perceptual block is the tendency to stereotype or see a solution to a problem that one is biased to see. This occurs because we have used similar techniques for solving the problem in the past. For example, if you have used a microcontroller to solve a certain type of problem, chances are that you are going to consider using a microcontroller in all related problems in the future. Another perceptual block is the difficulty of isolating the true problem. Three pictures that illustrate this are shown in Figure 4.2. When examining these images, people tend to form a conclusion as to what the content of each one is. Look carefully, as each picture has two equally valid interpretations.

Figure 4.2 Each of the images shown above has two different interpretations. Can you determine what they are?

One of the most common *emotional blocks* is the fear of failure. People often have creative ideas, but are afraid to express them since they may be criticized or may not have the "correct" answer. It is cliché to hear that you must fail often to succeed, but true. The highly successful product design company IDEO that was examined in Chapter 2 takes the approach in concept generation to "fail early and often" in order to succeed [Kel01]. Their design teams are encouraged to develop many seemingly outlandish ideas that are often discarded, but sometimes lead to innovative solutions. Another emotional block is a fear of chaos and disorganization. The creative process challenges engineers, since it is disorganized and not a neat scientific approach to which they are accustomed. Another block is the tendency to critically judge ideas, rather than generate and build upon them. Finally, it takes time for creative ideas to incubate. Most of us can relate to the experience when we could not solve a problem that nagged us for a period of time, followed by that unexpected "Aha!" moment when we identified the solution.

Environmental blocks refer to those things in our environment that limit creative ability. This could be in the form of poor teamwork where members distrust each other and criticize each other's ideas. In the workplace, this could be due to autocratic management that resists new ideas. There are also cultural biases against creativity. There is a bias against creativity as an approach to problem solving in the engineering field. This is usually based upon the reasoning that there is a single correct solution to a problem and creativity is an excuse for poor engineering. It is true that creativity and brainstorming alone do not solve engineering problems—the concepts generated need to be scrutinized using engineering principles to become viable innovations.

The final block that Adams identified is that of *intellectual and expressive*. In an engineering context, this means that the designer needs to have an understanding of intellectual tools that are applied to solve problems. For example, mathematics is a universal language for expressing and solving scientific problems. Specific examples in electrical and computer engineering are languages that describe the characteristics of systems such as functional, logical, and state behaviors. Examples in digital design are truth tables (input, output behavior) and state diagrams (stimulus-response). In the domain of electronics design, a functional approach (input, output, and function) is commonly used. Chapters 5 and 6 present tools for modeling the behavior of computer and electrical systems.

4.1.2 Vertical and Lateral Thinking

Edward DeBono is the father of a field known as **lateral thinking**, which offers a different perspective on the barriers to creativity. Lateral thinking is contrasted to what is known as the vertical thinking process [Deb67, Deb70]. Engineers tend to be vertical (or convergent) thinkers, meaning that they are good at taking a problem and proceeding logically to the solution. This is typically a sequential linear process, where the engineer starts at the highest level and successively refines elements of the design to solve the problem. This is usually based upon experience solving similar problems and conventional tools that are employed in that particular area.

The objective of lateral (or divergent) thinking is to identify creative solutions. It is not concerned with developing the solution for the problem, or right or wrong solutions. It encourages jumping around between ideas. In the words of DeBono "The vertical thinker says: 'I know what I am looking for.' The lateral thinker says: 'I am looking but I won't know what I am looking for until I have found it.'" The field of lateral thinking is characterized by puzzles of the following type found at Paul Sloane's Lateral Thinking Puzzles website (*http://dspace.dial.pipex.com/sloane*):

> *A body is discovered in a park in Chicago in the middle of summer. It has a fractured skull and many other broken bones, but the cause of death was hypothermia.*
>
> *A hunter aimed his gun carefully and fired. Seconds later, he realized his mistake. Minutes later, he was dead.*
>
> *A man is returning from Switzerland by train. If he had been in a nonsmoking car he would have died.*

The objective in these puzzles is to develop plausible scenarios that explain how each of the above situations could have happened. A solution for the first example is that a person stowed away in the wheel compartment of a jet airliner. While in flight he froze and died of hypothermia. When the plane prepared to land, it lowered its landing gear, causing the body to fall to the park, fracturing his skull, and breaking his bones. Can you develop plausible scenarios that describe each of them?

Vertical thinking focuses on sequential steps toward a solution and tries to determine the correctness of the solution throughout the process. This is very different from lateral thinking where there is nonlinear jumping around between steps and there is no attempt to discern between right and wrong. As such, lateral thinking is more apt to follow least likely paths to a solution, whereas vertical thinking follows the most likely paths. The goal in lateral thinking is to develop as many solutions as possible, while vertical thinking tries to narrow to a single solution.

Lateral thinking is appropriate for the concept generation phase. So should concept generation and brainstorming be done by the individual or by a team? DeBono and Osborn [Osb63] conclude that creativity is more effective by individuals than by teams. However, Osborn also points out that there is great value in applying creativity in teams, since it provides a place for the team to work together on problems and see other perspectives. Our anecdotal observations of student design teams supports this—group brainstorming is effective for developing concepts, new product ideas, new features, and different ways to combine technologies. This is because in groups, ideas are readily built upon by other team members. More mathematical, technical, and theoretical breakthroughs tend to be the work of the lone genius. Examples of this are the theory of relativity (Einstein), Boolean logic (Bool), and Shannon's sampling theorem. We have also observed that novice designers, who do not have much experience in concept generation, can benefit greatly from group brainstorming techniques.

4.1.3 Strategies to Enhance Creativity

There are valuable strategies that can be employed to enhance the creative process. The body of research on the subject is very large and key points are summarized as follows:

- *Have a questioning attitude.* One of the keys is to have a questioning attitude and challenge assumptions. The willingness to do this generally decreases as people age. Young children are highly creative and are constantly questioning everything, with questions such as "Why do trees have leaves?" or "Why is the sky blue?" Asking basic questions stimulates creativity and is applicable to technical designs. When examining a design with a microcontroller, ask questions such as "Is there a way to replace the microcontroller?," "Are there other features that I can achieve with the microcontroller?," and "Is there a better microcontroller that can be used?"

- *Practice being creative.* Research shows that people can improve their creative ability through conscious effort. For example, try solving the puzzles presented in this chapter and in the end of the chapter problems. Be conscious of things that bother you ("pet peeves") in your everyday life and try to develop new solutions for them.

- *Suspend judgment.* It is easy to criticize and immediately dismiss ideas, so it is important to defer judgment and be flexible in thinking. Seemingly outlandish ideas can lead to other concepts that are valuable solutions. The opportunity for new solutions is curtailed if ideas are immediately judged and discarded. Creative concepts can be developed by taking a concept and modifying it or combining it with other seemingly unrelated concepts.

- *Allow time.* The creative process needs time for incubation. The human mind needs time to work on problems, so set aside time to reflect on the problem and to allow it to incubate so that the "Aha!" moment of discovery can happen.
- *Think like a beginner.* New solutions often come from novices. The reason is that novices don't have preconceived ideas as to the solution for a problem. Experience is a double-edged sword—it allows one to quickly solve problems by drawing upon pre-existing solutions, but can inhibit creativity. If confronted with a new problem that bears similarity to one encountered in the past, then it is likely that the new solution will bear similarity to the old one. If everyone else is it doing it one way, consider the opposite.

Many creative ideas arise from novel combinations and adaptations of existing technology. SCAMPER, an acronym for substitute, combine, adapt, modify, put to other use, eliminate, and rearrange/reverse, can be used as a guide to systematically generate creative concepts. The SCAMPER principles are valuable in brainstorming and are described below:

- *Substitute.* Can new elements be substituted for those that already exist in the system?
- *Combine.* Can existing entities be combined in a novel way that has not been done before?
- *Adapt.* Can parts of the whole be adapted to operate differently?
- *Modify.* Can part or all of a system be modified? For example, size, shape, or functionality.
- *Put to other use.* Are there other application domains where the product or system can be put to use?
- *Eliminate.* Can parts of the whole be eliminated? Or should the whole itself be eliminated?
- *Rearrange or Reverse.* Can elements of the system be rearranged differently to work better? This is different from substituting in that the elements of the system are not changed, but rearranged or ordered differently to create something new. In terms of reversal, are there any roles or objectives that can be reversed?

SCAMPER is a modification of a set of questions that was originally posed by Osborn [Osb63] and was modified to its form above by Michalko [Mic91].

4.2 Concept Generation

After the problem is defined, the next step is to explore concepts for the solution. It is unlikely that a design team will have reached this stage without some ideas for solving the problem, but it is important to fully explore the design space. Ulrich and Eppinger [Ulr03] identify the following phases of concept generation—search internally, search externally, and systematically explore. Each is considered in turn.

External searching was covered to a great extent in Chapters 2 and 3, which addressed conducting background research and benchmarking. Methods of external searching are:

- Conduct literature search.
- Search and review existing patents.
- Benchmark similar products.
- Interview experts.

Internal searching is done by the team members via methods such as brainstorming. The team members need have to have a common problem definition for this to be effective. Understanding the tradeoffs by using requirement analysis methods in Chapter 3 is also valuable, as overcoming tradeoffs leads to innovative solutions. Furthermore, the team should decompose larger problems into subproblems and then attack the subproblems individually. Chapter 5 addresses the process of problem decomposition.

Figure 4.3 Wally brainstorming. (Dilbert © Scott Adams / Dist. by United Feature Syndicate, Inc.)

The most well-known method of internal searching is **brainstorming**. Group brainstorming is effective for generating many concepts in a short period of time (Figure 4.3). Experienced design teams are known to generate hundreds of concepts in an hour. Traditional brainstorming is not highly-structured—though a facilitator helps—and employs five basic rules:

- No criticism or judgment of ideas.
- Wild ideas are encouraged.

- Quantity is stressed over quality.
- Build upon and modify the ideas of others.
- All ideas are recorded.

Many novice design teams struggle with unstructured brainstorming and more formalized approaches, such as brainwriting and the nominal group technique, can be of benefit. The steps of *brainwriting* are:

1) The team develops a common problem statement that is read out loud.
2) Each team member writes ideas down on a card and places it in the center of the table.
3) Other team members then take cards from the pile and use others' ideas to generate new ones or build upon them, keeping in mind the principles of SCAMPER. Alternatively, members can each generate an idea, write it on a card, and then pass it to another team member. Each member then builds upon the idea passed to them.

Brainwriting 6-3-5 is a variation where the objective is to have six people develop three ideas in 5 minutes. The optimal number of people for the exercise is thought to be six, although it is not necessary. Each person generates three ideas in 5 minutes, and clearly describes it, using sketches and written descriptions on paper. At the end of 5 minutes, each team member passes the ideas to another team member. The next person reviews the ideas of the teammate and adds three more by building on them, developing new ones, or ignoring as necessary. This process continues until all members have reviewed all papers.

In the ***nominal group technique*** (NGT) [Del71] each team member silently generates ideas that are reported out in a round-robin fashion so that all members have an opportunity to present ideas. Concepts are selected by a multivoting scheme with each member casting a predetermined number of votes for the ideas presented. The ideas are then ranked, discussed further, and voted upon again if necessary. The steps of NGT are as follows:

- *Read problem statement.* It should be read out loud by a team member (the facilitator).
- *Restate the problem.* Each person restates the problem in his or her own words to ensure that all members understand it.
- *Silently generate ideas.* All members silently generate ideas during a set period of time, typically 5–15 minutes.
- *Collect ideas in a round-robin fashion.* Each person presents one idea in turn until all ideas are exhausted. The facilitator should clarify ideas and all should be written where the entire team can view them.
- *Summarize and rephrase ideas.* Once the ideas are collected, the facilitator leads a discussion to clarify and rephrase the ideas. This ensures that the entire group is familiar with them. Related ideas can be grouped or merged together.

- *Vote.* Each person casts a predetermined number of votes, typically three to six, for the ideas presented. The outcome is a set of prioritized ideas that the team can further discuss and pursue.

To systematically generate concepts, the problem is decomposed into sub-functions and solutions are sought for the subfunctions. A *concept table*, demonstrated in Table 4.1, is a tool for identifying different combinations, arrangements, and substitutions. The table headings identify functions to be achieved in the design, while the entries in the corresponding column represent potential solutions. Novel products or solutions are generated by combining elements from each of the columns, which are identified in the table by circled elements. The solutions can be in the form of a single element selected from each column, or as in the example shown, multiple elements selected from each column.

Table 4.1 A concept table for generating ideas for a personal computing system. The potential solution is identified by the combination of circled elements.

User Interface	Display	Connectivity and Expansion	Power	Size
Keyboard	CRT	Serial & Parallel	Battery	Handheld, Fits in Pocket
Touchpad	Flat Panel	USB	AC Power	Notebook Size
Handwriting Recognition	Plasma	Wireless Ethernet	Solar Power	Wearable
Video	Heads-up Display	Wired Ethernet	Fuel Cell	Credit Card Size
Voice	LCD	PCMCIA	Thermal Transfer	Flexible in Shape
		Modem / Telephone		

Using the concepts circled in Table 4.1, one can imagine a personal computing system that has the following features:

1) Is wearable with different credit card size components placed on the body and in clothing to make it comfortable to use.
2) Is powered by a combination of solar cells, fuel cells, and from thermal heat generated by a person's body.
3) Has a microphone and camera integrated in the user's clothing for interface to the system, as well as a flexible foldable keyboard for typing that is stored in a pocket.

4) Has a heads-up display integrated with the user's eyeglasses or baseball hat.
5) Has a miniature earpiece microphone used for communication.

While the above example focused on novel combinations and substitutions, the concept table can also be used to examine the possibility of eliminating ideas. For example, the table inherently assumes that a display will be used. However, it should also be asked if it is absolutely necessary in the design.

Another example is shown in Table 4.2, where the objective is to identify design concepts for a temperature measurement and display device. There are three main elements to the proposed solution: the thermal sensing method, circuitry that converts the sensor information (temperature) to a voltage, and a display unit that converts the voltage to a displayed temperature. Note that the table implies a three-stage architecture, thus concepts are generated within that framework. There may be completely different architectures that are better.

Table 4.2 Concept table for a temperature measurement device.

Thermal Sensing	Conversion to Voltage	Display
Thermistor	Op Amp Design	Seven-Segment LEDs
RTD	Transistor Designs	LCD
Thermocouple		Analog Dial Indicator

A related tool is a *concept fan*, which is a graphical representation of design decisions and choices. An example concept fan for the temperature measuring device is shown in Figure 4.4. Design decisions are identified by circles; solutions are indicated by squares. In this example, more options are shown than in Table 4.2. Concepts are generated by selecting among the different solution blocks.

4.3 Concept Evaluation

The concepts generated are evaluated to determine which are the most promising to pursue. The designer should exercise engineering judgment and use the customer needs and technical factors to drive the decision. This process is shown in Figure 4.5, where the user needs, concepts, and engineering consideration serve as inputs to a decision process to rank the concepts. A point of caution—some of the methods presented generate numerical scores for comparing concepts, leading one to potentially believe that the quantitative results are infallible. Keep in mind that the inputs are based on qualitative and semiquantitative assessments and can be geared to select a preconceived notion of the solution. It is important to maintain flexibility of thinking, to challenge assumptions, and ultimately determine the best concept.

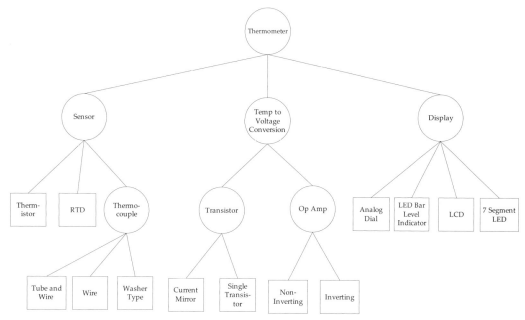

Figure 4.4 A concept fan for the temperature measuring device. The circles represent the choices to be made and the squares represent potential solutions to the choices.

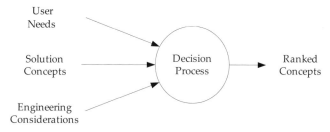

Figure 4.5 Process for concept evaluation.

4.3.1 Initial Evaluation

The concepts generated should be initially reviewed and those that are completely infeasible discarded. Some of the reasons a concept may be deemed infeasible are that it may be far too costly, will take too long to develop, or involve too much risk. In many cases it may be deemed that using cutting-edge technology represents an unacceptable risk. Concepts that clearly cannot meet the engineering requirements should also be discarded. Care should be taken not to completely eliminate ideas that may have merit, as conditions change and some concepts that were previously thought unrealistic may become viable in the future.

4.3.2 Strengths and Weaknesses Analysis

Another form of evaluation is to complete a *strengths and weaknesses analysis* of the potential solutions. Table 4.3 demonstrates the application of this analysis applied to an experimental design project for testing of the Intel 1000XF card [examined in Chapter 2 (Example 2.2) and Chapter 3 (Table 3.2)]. In order to test the card under different operating temperatures, a method of heating the card and holding its temperature fixed during the experiment was needed. The two solutions compared were to use a contact heating element and to place the card in an environmental test chamber. In this particular example, the temperature chamber solution was ultimately selected due to the need for a uniform temperature distribution. The strength and weakness analysis is good for examining problems of moderate complexity. It suffers in that it does not require uniform criteria for comparison. To make the method more quantitative, relative scores for the strengths (plus factors) and weaknesses (minus factors) can be assigned and used to score the concepts.

Table 4.3 A strengths and weaknesses analysis of proposed methods for heating an Intel 1000XF card to be used in lifetime testing [Ese03].

Method	Strengths	Weaknesses
Contact Heating	• Simplest design • Could be used internally to computer	• Does not create uniform temperature • Hard to control temperature
Temperature Chamber	• Uniform temperature • Greater control over temperature	• Must be external to computer • More difficult to design • Expensive

4.3.3 Analytical Hierarchy Process and Decision Matrices

In the Analytical Hierarchy Process (AHP), design alternatives are compared against preselected criteria, such as the engineering or marketing requirements. AHP is covered in detail in Appendix B and was first applied in Chapter 2 for project selection. The reader is encouraged to review Appendix B as necessary. The end result of AHP is a decision matrix as shown in Table 4.4, where the criteria are listed in the leftmost column with the associated weighting factors (ω_i) quantifying the relative importance of the criteria. The body of the matrix contains design ratings α_{ij} that reflect the technical merit of each of the jth design options relative to ith criterion. The total score S_j for each design option is computed as a weighted summation of the design ratings and weighting factors.

Table 4.4 A decision matrix for the Analytical Hierarchy Process.

		Design Option 1	Design Option 2	...	Design Option n
Criteria 1	ω_1	α_{11}	α_{12}	...	α_{1n}
Criteria 2	ω_2	α_{21}	α_{22}	...	α_{2n}
⋮	⋮	⋮	⋮	...	⋮
Criteria m	ω_m	α_{m1}	α_{m2}	...	α_{mn}
Score		$S_1 = \sum_{i=1}^{m} \omega_i \alpha_{i1}$	$S_2 = \sum_{i=1}^{m} \omega_i \alpha_{i2}$	...	$S_n = \sum_{i=1}^{m} \omega_i \alpha_{in}$

The application of AHP is demonstrated for the design of an electronic circuit for measuring temperature by producing a voltage signal that is directly proportional to temperature.

Step 1: Determine the Selection Criteria

Assume that the criteria for comparing the concepts are high accuracy, low cost, small size, and availability of parts for manufacture.

Step 2: Determine the Criteria Weightings

Assume that the criteria were ranked by pairwise comparison and weights computed (see Appendix B) as shown in Table 4.5.

Table 4.5 Pairwise comparison matrix.

	Accuracy	Cost	Size	Availability	Weights
Accuracy	1	5	3	1/4	0.42
Cost	1/5	1	2	1/4	0.12
Size	1/3	1/2	1	1	0.12
Availability	4	4	1	1	0.34

Step 3: Identify and Rate Alternatives Relative to the Criteria

Three candidate solutions are shown in Figure 4.6. Each acts as a constant current source that drives a temperature measurement device (resistance temperature detactor, or RTD). The resistance of an RTD varies with temperature, and when driven by a constant current I produces a voltage V_T that varies proportionally with temperature. Each circuit supplies a constant current of $I = 1$ mA.

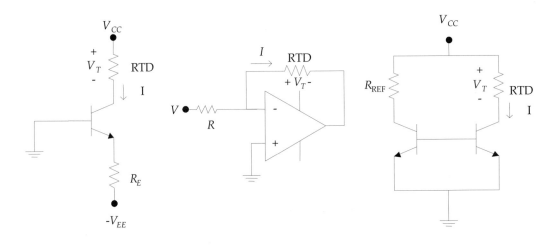

1: Single Transistor   2: Inverting Op Amp   3: Current Mirror

Figure 4.6 Candidate solutions for temperature measurement.

The accuracy of each design was evaluated by a sensitivity analysis, using a SPICE circuit simulation package assuming 10% resistors. The deviation of the output voltage (maximum deviation from nominal) for the three designs is 9.2%, 1.3%, and 1.9% respectively. Since the objective is to minimize the deviation, the following rating metric is used

$$\alpha = \frac{\min\{\text{deviation}\}}{\text{deviation}}. \tag{1}$$

This produces the following normalized design ratings for accuracy: $\alpha_{11} = 0.008$, $\alpha_{12} = 0.55$, and $\alpha_{13} = 0.37$.

The parts costs are the following: resistors = $0.05, bipolar junction transistors (BJTs) = $0.15, op amps = $0.35, and RTDs = $0.25. Using a measure for cost similar to (1) gives the following normalized cost ratings for the three options respectively: $\alpha_{21} = 0.41$, $\alpha_{22} = 0.28$, and $\alpha_{23} = 0.31$.

Assume that to manufacture each circuit on a printed circuit board requires the following dimensions: design 1 = 1 in², design 2 = 1.56 in², and design 3 = 2.25 in². The objective is to minimize size, and again using a measure analogous to (1) for the required space to manufacture each produces the following normalized decision ratings: $\alpha_{31} = 0.48$, $\alpha_{32} = 0.31$, and $\alpha_{33} = 0.21$.

Assume that the parts are available 95%, 70%, 90%, and 80% of the time for the resistors, BJTs, RTDs, and op amps respectively. A measure for the overall availability of parts to manufacture each design is required. One way to measure this is to compute the probability that a design will be able to be manufactured on the basis of the past history of part availability. This is found by multiplying the availability of all individual components needed for the design:

$$P(\text{design 1 can be produced}) = (0.95)(0.90)(0.70) = 0.60$$
$$P(\text{design 2 can be produced}) = (0.95)(0.90)(0.80) = 0.68$$
$$P(\text{design 3 can be produced}) = (0.95)(0.90)(0.70)(0.70) = 0.42$$

This produces the following normalized decision ratings for availability: $\alpha_{41} = 0.35$, $\alpha_{42} = 0.40$, and $\alpha_{43} = 0.25$.

Step 4: Compute Scores for the Alternatives

The decision matrix is built and the overall weighted scores for the alternatives are computed as shown in Table 4.6.

Table 4.6 The decision matrix.

		Single BJT	Op Amp	Current Mirror
Accuracy	0.42	0.08	0.55	0.37
Cost	0.12	0.41	0.28	0.31
Size	0.12	0.48	0.31	0.21
Availability	0.34	0.35	0.40	0.25
Score		0.26	0.44	0.30

Step 5: Review the Decision

Remember that this is a semiquantitative method. The final ranking indicates that design options 1 and 3 are quite similar, while both are inferior to option 2.

4.3.4 Pugh Concept Selection

Pugh concept selection is a method of comparing concepts against criteria, similar to what we saw with a decision matrix. It is different in that it has a simpler scoring method and it is an iterative process. The steps of Pugh concept selection are:

1) Select the comparison criteria, usually the engineering or marketing requirements.
2) Determine weights for the criteria.
3) Determine the concepts.

4) Select a baseline concept that is initially believed to be the best.
5) Compare all other concepts to the baseline, using the following scoring method: +1 better than, 0 equal to, −1 worse than.
6) Compute a weighted score for each concept, not including the baseline.
7) Examine each concept to determine if it should be retained, updated, or dropped. Synthesize the best elements of others into other concepts wherever possible.
8) Update the table and iterate until a superior concept emerges.

An example of a Pugh concept selection matrix is shown in Table 4.7.

Table 4.7 Pugh concept selection matrix.

		Option 1 (Reference)	Option 2	Option 3	Option 4
Criteria 1	4	-	0	0	+1
Criteria 2	5	-	+1	−1	0
Criteria 3	2	-	−1	0	+1
Criteria 4	1	-	+1	+1	−1
Score		-	4	−4	5
Continue?		Combine	Yes	No	Combine

4.4 Project Application: Concept Generation and Evaluation

The following advice is provided for teams in the concept generation and evaluation phase:

- Set aside time specifically for concept generation and evaluation and take it as a challenge to identify as many concepts as possible.
- Search externally, including literature reviews and patent searches.
- Search internally using brainstorming, brainwriting, or the nominal group technique. Effective teams generate many concepts in a brainstorming session.
- Examine solutions for the entire design, for subfunctions of the design, and for individual components (such as integrated circuit selection). The techniques in this chapter can be combined with design methods presented in Chapters 5 and 6.
- Utilize SCAMPER, concept tables, and concept fans as tools to facilitate and document concept generation.
- Critically and objectively evaluate concepts against common criteria.
- Clearly identify the concept(s) selected and the rationale for selection.

4.5 Summary and Further Reading

In the design process, it is important to creatively generate different concepts for a solution to a problem. This is followed by an evaluation of concepts to determine which are the most promising. This chapter identified barriers to creativity and provided strategies for enhancing creative ability. The concepts of vertical and lateral thinking were introduced, and their impact on the design process was explored. Methods of concept generation, including brainstorming, concept tables, and concept fans were presented. Finally, methods for critically evaluating concepts (strength/weakness analysis, Analytical Hierarchy Process, and Pugh concept selection) were presented.

There are many references that examine creativity and concept generation. Adams [Ada01] is a good reference for creativity and problem solving with a technical bent. Alex Osborn was an advocate of creativity and developed two readable works that address the creative process, the need for creativity, and strategies for enhancing it [Osb48, Osb63]. Edward DeBono is another well-known authority in the field and has produced many works on lateral thinking and creativity [Deb67, Deb70]. Paul Sloane has published numerous books with lateral thinking puzzles [Slo91, Slo93, Slo94]. TRIZ [Alt99] is a more advanced and complex approach to concept generation, that centers around resolving tradeoffs in a problem. It is fairly complex and may be considered by more advanced teams.

4.6 Problems

4.1 Consider the nine dot puzzle shown in Figure 4.1(b). Draw **three** connected straight lines that pass through all nine dots.

4.2 Consider the six sticks shown below. Rearrange the sticks to produce four equilateral triangles (the sticks cannot be broken).

4.3 Consider the fish shown below made of eight sticks and a coin for the eye. The objective is to make the fish face the other direction by moving only the coin and three sticks.

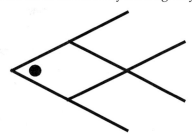

82 Design for Electrical and Computer Engineers

4.4 For each of the following lateral thinking puzzles, develop a plausible solution (from Paul Sloane's *Lateral Thinking Puzzles* [http://dspace.dial.pipex.com/sloane]):

   a) A man walks into a bar and asks the barman for a glass of water. The barman pulls out a gun and points it at the man. The man says "Thank you" and walks out.

   b) A woman had two sons who were born on the same hour of the same day of the same year. But they were not twins. How could this be so?

   c) Why is it better to have round manhole covers than square ones?

   d) A man went to a party and drank some of the punch. He then left early. Everyone else at the party who drank the punch subsequently died of poisoning. Why did the man not die?

4.5 Legislation was passed to allow handguns in the cockpits of passenger airliners to prevent hijacking. Brainstorm to develop concepts that prevent anyone other than the pilot from using the handgun.

4.6 Imagine if scientists and engineers were able to develop a technology that would allow people to be transported from any place on earth to another instantaneously. Brainstorm to determine the potential impact this would have on society.

4.7 Student advising at many colleges and universities is seen as an area that can be improved. Brainstorm to develop ideas as to how student advising could be improved at your college or university.

4.8 In your own words, describe what a concept table and a concept fan are.

4.9 Consider the problem solved in Section 4.3.3. For this example assume that:

   - The following is the result of the paired comparison.

	Accuracy	Cost	Size	Availability
Accuracy	1	1/3	2	1/2
Cost	3	1	5	1
Size	1/2	1/5	1	2
Availability	2	1	1/2	1

   - The parts costs are the following: resistors = $0.05, bipolar junction transistors (BJTs) = $0.10, op amps = $0.35, and RTDs = $0.25.

   - The parts are available 99%, 90%, 85%, and 70% of the time for the resistors, BJTs, RTDs, and op amps respectively.

   - Everything else is the same as presented in Section 4.3.3.

Compute the rankings of the design options, using a weighted decision matrix of the type shown in Table 4.6.

4.10 **Project Application.** Utilize the methods in this chapter to generate concepts for your particular design problem. Critically evaluate the concepts generated, using one or more of the techniques presented in the chapter that is appropriate for the problem. Section 4.4 provides guidance on how to conduct this process and document the results.

Part II — Design Tools

Chapter 5 System Design I: Functional Decomposition

At Sony, we assume all products of our competitors will have basically the same technology, price, performance, and features. Design is the one thing that differentiates one product from another in the marketplace. —Norio Ohgo, Chairman and CEO, Sony

After the technical concept is selected, it is translated into a solution that satisfies the system requirements. The designer must put on paper, or the computer screen, a representation that is meaningful and clear; in other words, a useful abstraction of the system. Engineering designs are often complex, consisting of many systems and subsystems, thus this representation should facilitate the design process and effectively describe the system. In addition, it serves an important function in communicating the design to all members of the team. Imagine a scenario where each team member is responsible for designing part of a large system. Each person develops a part in isolation and several months later the team gets back together to integrate the pieces. Of course, the system won't work unless the team has collectively defined and communicated the functionality and interfaces for all subsystems in the design.

This chapter presents a well-known design technique—known as *functional decomposition*—that is intuitive, flexible, and straightforward to apply. It is probably the most pervasive design technique used for engineering systems and is applicable to a wide variety of problems that extend well beyond electrical and computer engineering. In functional decomposition, systems are designed by determining the overall functionality and then iteratively decomposing it into component subsystems, each with its own functionality.

The objective of this chapter is to present both basic design concepts and the functional decomposition design technique. A process for functional decomposition is provided and it is applied to examples in analog electronics, digital electronics, and software systems.

Learning Objectives

By the end of this chapter, the reader should:
- Understand the differences between bottom-up and top-down design.
- Know what functional decomposition is and how to apply it.
- Be able to apply functional decomposition to different problem domains.
- Understand the concepts of coupling and cohesion, and how they impact designs.

5.1 Bottom-Up and Top-Down Design

Two general approaches to synthesizing engineering designs are known as *bottom-up* and *top-down*. In the case of bottom-up, the designer starts with basic components and synthesizes them to create the overall system. To use an analogy, consider the case of creating an automobile. In the bottom-up approach, you have pieces of the automobile, such as the tires, motor, axle, transmission, alternator, and they are brought together to create a car. The implication is that the final system depends upon the parts at hand. In other words, in the bottom-up approach, the parts and subsystems are given, and from them an artifact is created.

The top-down approach is analogous to the concept of divide and conquer. In top-down the designer has an overall vision of what the final system must do, and the problem is partitioned into components, or subsystems that work together to achieve the overall goal. Then each subsystem is successively refined and partitioned as necessary. In the case of the automobile, the overall objective is determined; the major subsystems are defined, such as electrical, power drive train, and the suspension; and then each subsystem is further refined into its component parts until the complete system is designed.

A debate that continues in the design community revolves around which is the better approach. It might appear that top-down is better, since it starts with the overall goal (requirements) and from that a solution is developed. Top-down is particularly valuable on large projects with many subsystems, where it is unlikely that bringing together pieces in an ad-hoc fashion will successfully solve the problem. A disadvantage of top-down design is that it tends to limit the solution space and innovation. Top-down design is inclined to follow a vertical thought process (Chapter 4) where the designer starts with a problem and successively refines the subsystems until a blueprint for solving the problem is defined. Furthermore, the designer cannot create a top-down design in a vacuum without bottom-up knowledge of existing technology and how the system can be realized.

Bottom-up has the advantage of lending itself to creativity. It allows the designer to take different technologies and from them create something new, allowing more "what if?" questions to be asked. Bottom-up design is applicable when there are constraints on the components that can be used. This is a realistic scenario. Consider the case of variant design, where the goal is to improve the performance of an existing, or legacy, system. For example, automobile manufacturers might have to redesign their models to meet new emissions, mileage, or safety standards. If you are not starting with a new design and must utilize existing systems, it requires bottom-up thinking. In reality, most problems require a combination of bottom-up and top-down thinking, and the designer must alternate between them.

In summary, it is most effective to work between bottom-up and top-down. A completely top-down approach is not feasible because the designer must have an understanding of the bottom level technology for the components of the design hierarchy to be realistic. Likewise, completely bottom-up by itself is generally not feasible, particularly as the system complexity grows.

5.2 Functional Decomposition

Functional decomposition is a recursive process that iteratively describes the functionality of all system components. It is analogous to the mathematical concept of a function, for example, $y = f(x)$. In this function there is an input x, an output y, and a transformation between the input and output $f()$. This is easily extended to the case of multiple inputs and outputs where the inputs and outputs are vectors, $\vec{y} = f(\vec{x})$. In functional decomposition, the same items are defined as in the mathematical analogy—the inputs, the outputs, and the transformation between the inputs and outputs (the functionality). Those three items constitute what is known as the *functional specification* or *functional requirement*—the requirement that a functional module should meet. A *module* is a block, or subsystem, that performs a function. Functional decomposition has a strong top-down flavor, due to the fact that the highest level functionality is defined and then further refined into subfunctions, each with its own inputs, outputs, and functionality. The process is repeated until some base level functionality is reached where the modules can be actualized with physical components.

A process for applying functional decomposition is illustrated in Figure 5.1. It starts with a definition of the highest level (Level 0) of system functionality (the functional requirement for the system). This is followed by definition of the next level of the hierarchy that is needed to achieve the design objective. The Level 1 design is typically referred to as the main *design architecture* of the system. In this context, architecture means the organization and interconnections between modules. Care must be taken at each design level to ensure that it satisfies the requirements of the higher level. The process is repeated for successive levels of the design and stops when the *detailed design* level is reached. Detailed design is where the problem can be decomposed no further and the identification of elements such as circuit components, logic gates, or software code takes place. The number of levels in the design hierarchy depends upon the complexity of the problem.

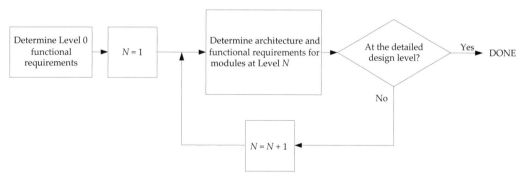

Figure 5.1 A process for developing designs using functional decomposition.

5.3 Guidance

The following guidance is provided before examining applications of the functional decomposition technique:

- *It is an iterative process*. During the first pass, it is not possible to know all of the detailed interfaces between components and the exact functionality of each block. In act, some details are not known until the implementation level is reached, so the designer needs to iterate, work between top-down and bottom-up, and adjust the design as necessary.
- *Set aside sufficient time to develop the design*. This is a corollary to the previous point. The iterative nature means that it takes time to examine different solutions and to refine the etails into a working solution.
- *Pair together items of similar complexity*. Modules at each level should have similar complexity and granularity.
- *A good design will have the interfaces and functionality of modules well-defined*. It is fairly easy to piece together some blocks into an apparently reasonable design. However, the functional requirements should be clearly defined and the technical feasibility understood. If not, the design will fall apart when it comes to the implementation stage. Consider the following advice of a well-known architectural designer:

    The details are not the details. They make the design. —Charles Eames

- *Look for innovations*. Top-down designs tend to follow a vertical thinking process, where the designer proceeds linearly from problem to solution. Try to incorporate lateral thinking strategies from Chapter 4 and examine alternative architectures and technologies for the solution.
- *Don't take functional requirements to the absurd level*. Common elements, such as analog multipliers or digital logic gates, do not require explicit functional specifications. Doing so may become cumbersome and add little to the design. However, it depends upon the level at which you are working. If the goal is to design an analog multiplier chip, it is entirely appropriate to develop the functional requirement for the multiplier.
- *Combine functional decomposition with other methods of describing system behavior*. There is no single method or unifying theory for developing designs. Functional decomposition alone cannot describe all system behaviors. It may be supplemented by other tools such as flowcharts (logical behavior), state diagrams (stimulus-response), or data flow diagrams. In the digital stopwatch example presented later in this chapter, the behavior is defined by state diagrams. Other methods for describing system behavior are addressed in Chapter 6.

- *Find similar design architectures.* Determine if there exist similar designs and how they operate. Realize that this creates a bias toward existing solutions.
- *Use existing technology.* Many designers take the attitude that they are going to develop the entire design themselves, the sentiment being to ignore technology that they did not develop. Furthermore, engineering education predisposes us to design at a fundamental level. Both factors lead to time spent reinventing the wheel. If existing technology is available that meets both the engineering and cost requirements, then use it.
- *Keep it simple.* Do not add complexity that is not needed.

  > *A designer knows that he has achieved perfection not when there is nothing left to add, but when there is nothing left to take away.* —Antoine de St-Exupery

- *Communicate the results.* It is important to describe the theory of operation (the *why*) as well as the implementation (the *what*). The *what* in the completed design is usually quite clear from the implementation, but documenting the description of operation and design decisions helps later when the system must upgraded. Designs can also become very complex, so consider how much information can be effectively communicated on a single page. If the information is too complex to show reasonably on a page or two, then it probably is too detailed and another level in the hierarchy should be added.

5.4 Application: Electronics Design

We now examine the application of functional decomposition in different problem domains. In the domain of analog electronics, the inputs and outputs of modules are voltage and current signals. Typical transformations applied to the inputs are alterations in amplitude, power, phase, frequency, and spectral characteristics. Consider the design of an audio power amplifier that has the following engineering requirements.

The system must

- Accept an audio input signal source with a maximum input voltage of 0.5 V peak.
- Have adjustable volume control between zero and the maximum volume level.
- Deliver a maximum of 50 W to an 8 Ω speaker.
- Be powered by a standard 120 V, 60 Hz, AC outlet.

Level 0

The Level 0 functionality for the amplifier is shown in Figure 5.2, which is fairly simple—the inputs are an audio signal, volume control, and wall outlet power, and the output is an amplified audio signal.

Figure 5.2 Level 0 audio power amplifier functionality.

The system should be described in as much detail as possible for each level via the functional requirement. The Level 0 functional requirement for this design is as follows.

Module	Audio Power Amplifier
Inputs	- Audio input signal: 0.5 V peak. - Power: 120 V AC rms, 60 Hz. - User volume control: variable control.
Outputs	- Audio output signal: ? V peak value.
Functionality	Amplify the input signal to produce a 50-W maximum output signal. The amplification should have variable user control. The output volume should be variable between no volume and a maximum volume level.

Not all values can be known on the first pass through the design, as was indicated in the guidelines. Underlined items represent values that need to be determined or refined as the design proceeds. In this case, the peak value of the audio output voltage is determined from the system requirements on power gain. Knowing that the maximum power is given by $P_{max} = V_{peak}^2 / R$ allows the maximum output voltage to be computed as $V_{peak} = \sqrt{8\,\Omega * 50\text{ W}} = 20\text{ V}$.

Level 1

The Level 1 diagram, or system architecture, is shown in Figure 5.3. This architecture is common in amplifier design and is but one possible solution. It contains three cascaded amplifier stages and a DC supply that powers the three stages. The first amplifier stage, the *buffer amplifier*, provides a high-resistance buffer that minimizes loading effects with the source. Buffer amplifiers have extremely high input resistance and a unity signal gain. The *high-gain amplifier* increases the amplitude of the signal, but provides little in terms of the output current necessary to drive the speakers. The last stage in the cascade is the *power output stage*, which provides the current needed to drive the speakers, but has no voltage amplification.

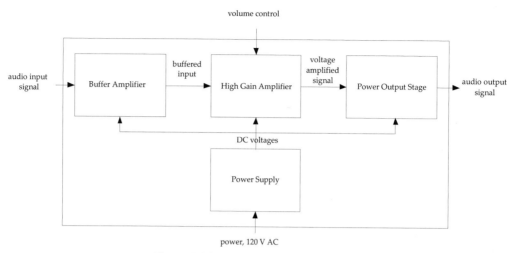

Figure 5.3 Level 1 audio amplifier design.

The functional requirements for the Level 1 subsystems are now detailed, starting with the buffer amplifier.

Module	Buffer amplifier
Inputs	- Audio input signal: 0.5 V peak. - Power: ± 25 V DC.
Outputs	- Audio signal: 0.5 V peak.
Functionality	Buffer the input signal and provide unity voltage gain. It should have an input resistance > 1 MΩ and an output resistance < 100 Ω.

Where did the ±25 V DC value for the DC input power come from? The system must produce a ±20 V AC output signal to satisfy the Level 0 requirement, so supply values that exceed that are required to power the electronics. How about the values for the input and output resistance? They are educated guesses, based on knowledge of what is achievable with the technology (bottom-up knowledge). The exact resistance requirements are refined later on the basis of the overall design, taking into account the input and output resistances for all stages.

Now consider the functional requirements for the high-gain amplifier.

Module	High-gain amplifier
Inputs	- Audio input signal: 0.5 V peak. - User volume control: variable control. - Power: ± 25 V DC
Outputs	- Audio signal: 20 V peak.
Functionality	Provide an adjustable voltage gain, between 1 and 40. It should have an input resistance > 100 kΩ and an output resistance < 100 Ω.

The gain of 40 is determined from the overall system power and gain requirements (the maximum input voltage of 0.5 V must be able to be amplified to 20 V), while the resistances are again educated guesses.

Now consider the power output stage.

Module	Power Output Stage
Inputs	- Audio input signal: 20 V peak. - Power: ± 25 V DC.
Outputs	- Audio signal: 20 V peak at up to 2.5 A.
Functionality	Provide unity voltage gain with output current as required by a resistive load of up to 2.5 A. It should have an input resistance > 1 MΩ and output resistance < 1 Ω.

For the power output stage, it is clear that 20 V peak needs to be delivered, but how was the requirement on current determined? The current needed to drive the speaker is determined from Ohm's law as $I = V/R = 20/8 = 2.5$ A.

The last module to examine at this level is the power supply.

Module	Power Supply
Inputs	- 120 V AC rms.
Outputs	- Power: ± 25 V DC with up to 3.0 A of current with a regulation of < 1%.
Functionality	Convert AC wall outlet voltage to positive and negative DC output voltages, and provide enough current to drive all amplifiers.

It is clear that the power supply needs to deliver ± 25 V DC, while the 3.0 A current capability was selected to supply the 2.5 A needed for the peak output power requirement plus the current needed to power the other amplifier stages.

Finally, it is necessary to determine if the values of the input and output resistances selected for the stages are realistic. For cascaded amplifier stages, the overall voltage gain is given by the product of gains multiplied by the voltage divider losses between stages [Sed04]. In this case the overall gain is

$$\text{Voltage gain} = \text{gain}_1 \times \text{gain}_2 \times \text{gain}_3 \left(\frac{R_{in2}}{R_{in2} + R_{out1}} \right) \left(\frac{R_{in3}}{R_{in3} + R_{out2}} \right)$$

$$= 1 \times 40 \times 1 \left(\frac{100\text{ k}}{100\text{ k} + 100} \right) \left(\frac{1\text{M}}{1\text{M} + 100} \right) \quad (1)$$

$$\approx 40$$

So the resistance values selected in the functional requirements satisfy the overall system requirements. If not, it would be necessary to go back and refine them.

Level 2

At this point, the three amplifier stages are ready for detailed component level design, while the power supply needs another level of refinement, as shown in Figure 5.4. The functional requirement for each of the elements in the power supply would be developed similarly. Functional decomposition stops at this point—all levels of the hierarchy are defined and the next step is the detailed design, where the actual circuit components are determined.

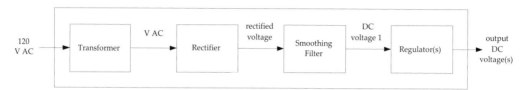

Figure 5.4 Level 2 design of the power supply.

5.5 Application: Digital Design

Functional decomposition is widely applied to the design of digital systems, where it is known as *entity-architecture* design. The inputs and outputs refer to the entity, and the architecture describes the functionality. The application of functional decomposition to digital systems is demonstrated in the following example. Consider the design of a simple digital stopwatch that keeps track of seconds and has the following engineering requirements.

The system must

- Have no more than two control buttons.
- Implement run, stop, and reset functions.
- Output a 16-bit binary number that represents seconds elapsed.

Level 0

The Level 0 diagram and functional requirements are shown in Figure 5.5.

Figure 5.5 Level 0 digital stopwatch functionality.

Module	Stopwatch
Inputs	- A: Reset button signal. When the button is pushed it resets the counter to zero. - B: Run/stop toggle signal. When the button is pushed it toggles between run and stop modes.
Outputs	- b_{15}–b_0: 16-bit binary number that represents the number of seconds elapsed.
Functionality	The stopwatch counts the number of seconds after B is pushed when the system is in the reset or stop mode. When in run mode and B is pushed, the stopwatch stops counting. A reset button push (A) will reset the output value of the counter to zero only when the stopwatch is in stop mode.

Level 1

The Level 1 architecture in Figure 5.6 contains three modules: a seconds counter, a clock divider, and a finite state machine (FSM). The stopwatch counts seconds, thus the seconds counter module counts the seconds and outputs a 16-bit number representing the number of seconds elapsed. The clock divider generates a 1 Hz signal that triggers the seconds counter. The FSM responds to the button press stimuli and produces the appropriate control signals for the seconds counter. The system clock is included to clock both the FSM and the clock divider.

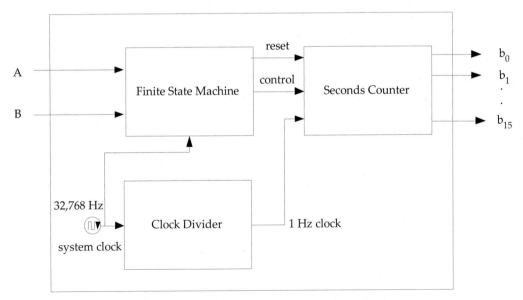

Figure 5.6 Level 1 design for the digital stopwatch.

The functionality of the Level 1 modules is described as follows, starting with the finite state machine.

Module	Finite State Machine
Inputs	- A: Signal to reset the counter. - B: Signal to toggle the stopwatch between run and stop modes. - Clock: 1 Hz clock signal.
Outputs	- Reset: Signal to reset the counter to zero. - Control: Signal that enables or disables the counter.
Functionality	State diagram with states Reset, Run, and Stop. Reset → Run on B; Run → Stop on B; Stop → Run on B; Stop → Reset on A.

The functionality of the finite state machine is described with a tool that is probably familiar to the reader, the state diagram. State diagrams are covered in more detail in Chapter 6. The state diagram describes stimulus-response behavior, and shows how the system transitions between states according to logic signals from the button presses.

Next, consider the clock divider.

Module	Clock Divider
Inputs	- System clock: <u>32,768</u> Hz.
Outputs	- Internal clock: 1 Hz clock for seconds elapsed.
Functionality	Divide the system clock by 32,768 to produce a 1 Hz clock.

The value of 32,768 Hz was selected for the system clock for several reasons. It is a power of 2 that is easily divisible by digital circuitry to produce a 1 Hz output signal. It is also well above the clock rate needed for detecting button presses, and there is a wide selection of crystals that can meet this requirement.

Finally, consider the seconds counter.

Module	Seconds Counter
Inputs	- Reset: Reset the counter to zero. - Control: Enable/disable the counter. - Clock: Increment the counter.
Outputs	- b_{15}–b_0: 16-bit binary representation of number of seconds elapsed.
Functionality	Count the seconds when enabled and resets to zero when reset signal enabled.

The system decomposition would end here, assuming that the design is to be implemented with off-the-shelf chips. The next step would be to determine components at the detailed design level. However, if it were an integrated circuit design, the description would continue until the transistor level is reached.

5.6 Application: Software Design

Software also lends itself to functional decomposition, since virtually all computing languages provide the capability to call functions, subroutines, or modules. Functional software design simplifies program development by eliminating the need to create redundant code via the use of functions that are called repeatedly.

Structure charts are specialized block diagrams for visualizing functional software designs. The modules used in a structure chart are shown in Figure 5.7. The larger arrows indicate connections to other modules, while the smaller arrows represent data and control information passed between modules. Five basic modules are utilized:

1) *Input modules*. Receive information.
2) *Output modules*. Return information.
3) *Transform modules*. Receive information, change it, and return the changed information.
4) *Coordination modules*. Coordinate or synchronize activities between modules.
5) *Composite modules*. Any possible combination of the other four.

This approach to software design, also known as *structured design*, was formalized in the 1970s by IBM researchers [Ste99].

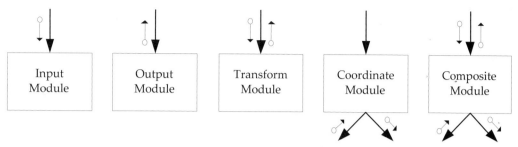

Figure 5.7 Module types for functional software design. The larger arrows indicate connections between modules and the smaller arrows represent data and control.

The following example demonstrates the application of functional decomposition to a software design with the following requirements.

The system must

- Accept an ASCII file of integer numbers as input.
- Sort the numbers into ascending order and save the sorted numbers to disk.
- Compute the mean of the numbers.
- Display the mean on the screen.

This is a fairly simple task that could easily be done in a single function, but doing so would not allow components of the design to be easily reused, tested, or troubleshot. The engineering requirements themselves provide some guidance in terms of how to arrange the functionality of the modules (*form follows function*). The architecture in Figure 5.8 contains a `main` module that calls three submodules. In this design `main` is a coordinating module that controls the processing and calling of the other modules, a common scenario. It was also decided that all user interaction would take place within `main`. The order of the processing is not described by structure charts. In our program, `main` calls `ReadArray`, `SortArray`, and `ComputeMean` in sequential order. `main` passes the filename (`fname`) to `ReadArray`, which reads in the array and the number of elements in it, and returns this information to `main`. The choice of passing in the filename was deliberate; the user could have been prompted for the filename in `ReadArray`, but doing so might limit future reuse of the function since you may not always want to do so when reading an array of data. `SortArray` is then called, which accepts the array of numbers and the number of elements in the array, and returns the sorted values in the same array. Finally, `ComputeMean` is executed, which accepts the sorted array and the number of elements, computes the mean value, and returns it to `main`.

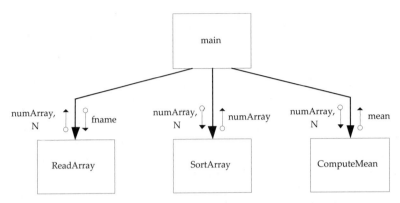

Figure 5.8 Structure chart design of sorting and mean computation program.

The functional requirements for each module in the structure chart are detailed in Table 5.1. The structure chart provides a visual relationship between modules in the design, but also has some disadvantages. It is difficult to visualize designs as the complexity of the software increases. This can be addressed by expanding sublevels in the design as necessary in different diagrams. Structure charts also lack a temporal aspect that indicates the calling order. Most software systems have many layers in the hierarchy and highly complex calling patterns. In this example, main calls three modules in a well-defined order, but if there were another level in the hierarchy, there is no reason why it could not be called by a module at any other level. That leads to some of the unique problems associated with software design. Functional design works well for small to moderately complex software, but tends to fall short when applied to large-scale software systems. As such, it has given way to the object-oriented design approach.

5.7 Application: Thermometer Design

The final example includes both analog and digital modules and the objective is to design a thermometer that meets the following engineering requirements.

The system must
- Measure temperature between 0 and 200°C.
- Have an accuracy of 0.4% of full scale.
- Display the temperature digitally, including one digit beyond the decimal point.
- Be powered by a standard 120 V, 60 Hz AC outlet.
- Use an RTD (resistance temperature detector) that has an accuracy of 0.55°C over the range. The resistance of the RTD varies linearly with temperature from 100 Ω at 0°C to 178 Ω at 200°C. (Note: this requirement does not meet the abstractness property identified in Chapter 3, since it identifies part of the solution. This requirement is given to provide guidance in this example.)

Table 5.1 Functional design requirements for the number sort program.

Module name	main()
Module type	Coordination
Input arguments	None.
Output arguments	None.
Description	The main function calls ReadArray() to read the input file from disk, SortArray() to sort the array, and ComputeMean() to determine the mean value of elements in the array. User interaction requires the user to enter the filename, and the mean value is displayed on the screen.
Modules invoked	ReadArray, SortArray, and ComputeMean.

Module name	ReadArray()
Module type	Input and output
Input arguments	- fname[]: character array with filename to read from.
Output Arguments	- numArray[]: integer array with elements read from file. - N: number of elements in numArray[].
Description	Read data from input data file and store elements in array numArray[]. The number of elements read is placed in N.
Modules invoked	None.

Module name	SortArray()
Module type	Transformation
Input arguments	- numArray[]: integer array of numbers. - N: number of elements in numArray[].
Output Arguments	- numArray[]: sorted array of integer numbers.
Description	Sort elements in array using a shell sort algorithm. Saves the sorted array to disk.
Modules invoked	None.

Module name	ComputeMean()
Module type	Input and output
Input arguments	- numArray[]: integer array of numbers. - N: number of elements in numArray[].
Output arguments	- mean: mean value of the elements in the array.
Description	Computes the mean value of the integer elements in the array.
Modules invoked	None.

Level 0

The overall goal is to convert a sensed temperature to a digital temperature reading. The Level 0 description is shown in Figure 5.9.

Figure 5.9 Level 0 digital thermometer functionality.

Module	Digital Thermometer
Inputs	- Ambient temperature: 0–200°C. - Power: 120 V AC power.
Outputs	- Digital temperature display: A four digit display, including one digit beyond the decimal point.
Functionality	Displays temperature on digital readout with an accuracy of 0.4% of full scale.

Level 1

The Level 1 architecture selected is shown in Figure 5.10. The temperature conversion unit converts the temperature to an analog voltage, using the RTD that is sampled by the analog-to-digital converter. The N-bit binary output from the converter is translated into binary-coded decimal (BCD). BCD is a 4-bit representation of the digits between 0 and 9. Since there are four display digits, there are four separate binary encoded outputs from the BCD conversion unit. Common seven-segment LEDs are used for the display. However, they do not directly accept BCD and instead have seven input lines, each of which is individually switched to control the display segments. The requirements did not specifically address cost or size constraints, nor clearly define the environment, so there are many possible solutions. For example, an analog-to-digital current converter could be used, integrated circuit temperature-sensing packages could be considered, and microcontroller-based solutions are feasible as well.

From a system design perspective, an error budget is needed to identify the maximum error that each subsystem may introduce, while still achieving the overall accuracy. In this case, error is introduced in the temperature conversion unit and A/D converter, but not in the remaining digital components. The overall accuracy that the system must achieve is 0.4%, and that translates into 0.8°C of allowable error for the 200°C range. Let's now examine the modules.

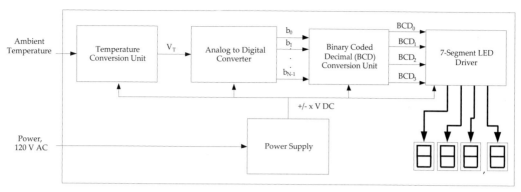

Figure 5.10 Level 1 design of the digital thermometer.

The functionality of the Level 1 modules is described as follows, starting with the temperature conversion unit.

Module	Temperature Conversion Unit
Inputs	- Ambient temperature: 0–200°C. - Power: ?_V DC (to power the electronics).
Outputs	- V_T: temperature proportional voltage. $V_T = \alpha T$, and ranges from ?_ to ?_ V.
Functionality	Produces an output voltage that is linearly proportional to temperature. It must achieve an accuracy of ?_%.

There are several unknowns at this point. The voltage necessary to power the electronics is not known, but a reasonable assumption could be made. The output voltage range and the accuracy are unknown. It is known that the RTD will introduce up to 0.55°C of error and that the electronics themselves will introduce additional error (the exact amount is unknown at this point). An educated guess is made that the maximum error allowed for the temperature unit is 0.6°C. This means that the electronics themselves would be required to introduce no more than 0.05°C of error as a result of the 0.55°C of error introduced by the RTD.

Now consider the analog to digital converter.

Module	A/D Converter
Inputs	- V_T: voltage proportional to temperature that ranges from ?_ to ?_ V. - Power: ?_ V DC.
Outputs	- b_{N-1}–b_0: ?_-bit binary representation of V_T.
Functionality	Converts analog input to binary digital output.

It is not likely that it will be necessary to design the A/D converter because low-cost, off-the-shelf solutions are available. The requirements drive the converter selection. There are two unknowns—the number of bits and the range of the input voltage. The number of bits affects the accuracy, since the greater the number of bits, the better the accuracy. The number of bits needed for the converter is calculated from the maximum allowable error that the A/D can introduce (0.2°C), the number of discrete intervals, and the temperate range as

$$\text{Max error} = \frac{\text{range}}{\text{number of intervals}} = \frac{200°C}{2^N} \leq 0.2°C \quad \Rightarrow \quad N \geq 9.97 \text{ bits.} \quad (2)$$

So the A/D converter needs to have at least 10 bits. How is the voltage range selected? It is typically fixed for a particular integrated circuit solution, but the temperature conversion subsystem output should be matched to the voltage range so that all bits are effectively utilized; otherwise, error is introduced.

Now, consider the BCD conversion unit.

Module	BCD Conversion Unit
Inputs	- 10-bit binary number (b_9–b_0): Represents the range 0.0–200.0°C. - Power: ? V DC.
Outputs	- BCD_0: 4-bit BCD representation of tenths digit (after decimal). - BCD_1: 4-bit BCD representation of ones digit. - BCD_2: 4-bit BCD representation of tens digit. - BCD_3: 4-bit BCD representation of hundreds digit.
Functionality	Converts the 10-bit binary number to BCD representation of temperature. Must refresh the displays twice a second.

The objective of the BCD conversion unit is fairly simple, although the component level design of the circuitry to accomplish the conversion is not.

This leads to the last module, the seven-segment LED driver, whose functionality is described as follows.

Module	Seven-Segment LED Driver
Inputs	- BCD_0: 4-bit BCD representation of tenths digit (after decimal). - BCD_1: 4-bit BCD representation of ones digit. - BCD_2: 4-bit BCD representation of tens digit. - BCD_3: 4-bit BCD representation of hundreds digit. - Power: ? V DC.
Outputs	- Four 7-segment driver lines.
Functionality	Converts the BCD for each digit into outputs that turn on LEDs in seven-segment package to display the temperature.

For completeness, the functional requirements of the power supply are supplied. They are similar to the power supply requirements utilized in the audio amplifier design in Section 5.4.

Module	Power supply
Inputs	- 120 V AC rms.
Outputs	- ± ? V DC with up to ? mA of current. - Regulation of ?%.
Functionality	Convert AC wall outlet voltage to positive and negative DC output voltages, with enough current to drive all circuit subsystems.

At this point, the requirements for the major subsystems are completed and ready for design at the component level. Illustration of the complete design would require a fair amount of detail, and while it is not presented here, some of the issues involved are discussed. First, there are a variety of electronic circuits (inverting op amps, single BJT configurations, and current mirrors, etc. — see Section 4.3.3) that could be utilized as a current source to drive the RTD in the temperature conversion subsystem. A midrange resolution A/D converter is needed, and its particular input voltage range drives the output voltage requirements for the temperature conversion module. The BCD conversion circuitry could be implemented using combinational digital logic (tedious due to the number of discrete gates), or a more efficient, but slower, sequential logic design. Finally, the seven-segment display converters could be designed using combinational logic that maps the BCD inputs into outputs to activate the appropriate display segments.

5.8 Coupling and Cohesion

The concepts of coupling and cohesion are examined before concluding this chapter. They originated to describe software designs [Ste99], but are applicable to electrical and computer systems. To understand their importance, consider the relationship between the number of modules in a system and the number of connections between them. For our purposes, a connection between two modules may consist of any number of signals without regard to their direction. Thus, a system consisting of two modules has, at most, one connection. If the number of modules is increased to three, the number of possible connections increases to three, a system with four modules has six possible connections, and five modules increases the number of possible connections to ten. The point is that the maximum number of potential connections increases rapidly with the number of modules in the system. The relationship between the maximum possible connections and number of modules (n) is given by

$$\text{Connections}_{max} = \frac{n(n-1)}{2}. \tag{3}$$

Modules are coupled if they depend upon each other in some way to operate properly. *Coupling* is the extent to which modules or subsystems are connected [Jal97]. Although there is no agreed upon mathematical definition of coupling, it seems obvious that increasing the exchange of control and data between two modules leads to a higher degree of coupling. When systems are highly coupled, it is difficult to change one module without affecting the other. Consider the extreme case where all modules in a system are connected to each other—an error in one module has the potential to affect every other module in the system. Errors in a module are propagated to others to a degree that is related to the amount of coupling. From this point of view, it is good to minimize coupling. Yet coupling cannot be eliminated, since the point of functional decomposition is to break a design into components that work together to produce a higher level behavior.

There are two ways to reduce coupling—minimize the number of connections between modules and maximize cohesion within modules. *Cohesion* refers to how focused a module is—highly cohesive systems do one or a few things very well. Stevens et al. [Ste99] defined six types of cohesion from the weakest to strongest as: coincidental, logical, temporal, communicational, sequential, and functional. More information on this can be found in the original work, but the conclusion is that modules with high functional cohesion are the most desirable. So it is best to design modules with a single well-defined functional objective consistent with the philosophy of functional decomposition. This leads to the important design principle that it is desirable to maximize cohesion, while minimizing coupling.

Coupling and cohesion impact the later stages of testing and system integration. If a particular module is highly cohesive, then it should be possible to test it independently of the other modules to verify its operability. This does not mean that it will necessarily operate properly when integrated into the overall system, but the probability that it will is higher if provided with proper inputs from connected modules. Contrast this to the case of a low-cohesion system. In that case, it will likely be difficult to test the individual modules without first integrating them.

To develop a better understanding, consider the amplifier design in Figure 5.3 (Section 5.4) with three cascaded amplifier stages. Each stage is highly cohesive, performing a singular function of signal amplification. Each of these stages could easily operate as a stand-alone module independent of the complete system. How about coupling? In terms of the number of connections, it is fairly low as each amplifier stage has an input and output voltage signal. The most coupled module in the system is the power supply, and not surprisingly, its failure leads to a complete system failure. Coupling in this case can also be viewed in terms of the resistance matching between input and output of the cascaded stages, producing the voltage divider effect in equation (1). For voltage amplifiers, the goal is to have high input resistance and low output resistance, which minimize both voltage losses and coupling. The stages are not completely uncoupled, because the input resistances, although large, are not infinite, and the output resistances are not zero. The modules in the power supply unit in Figure 5.4 (rectifier, smoothing filter, and regulator) have a much higher degree of coupling. In fact, it is difficult to develop a clear functional decomposition of the power supply module because the elements in the smoothing filter also serve as part of the rectifier circuit (refer to a basic electronics textbook [Sed04] for more information).

As another example, consider a software design where two options are under consideration: one large function with 1000 lines of code, versus 15 cohesive functions, each with an average of 100 lines of code. Both perform the same function, but which runs faster? Most likely the first, as it would be highly integrated and would not suffer from overhead needed with multiple functions. Which is easier to upgrade and debug a year from now? That is clearly the second case. Although loosely coupled and highly cohesive designs may facilitate better design and testing, they may not be best in terms of performance.

5.9 Project Application: The Functional Design

The following is a format for documenting and presenting functional designs.

Design Level 0

- Present a single module block diagram with inputs and outputs identified.
- Present the functional requirements: inputs, outputs, and functionality.

Design Level 1

- Present the Level 1 diagram (system architecture) with all modules and interconnections shown.
- Describe the theory of operation. This should explain how the modules work together to achieve the functional objectives.
- Present the functional requirements for each module at this level.

Design Level N (for $N > 1$)

- Repeat the process from Design Level 1 for as many levels as necessary.

Design Alternatives

- Describe the different alternatives that were considered, the tradeoffs, and the rationale for the choices made. This should be based upon concept evaluation methods communicated in Chapter 4.

5.10 Summary and Further Reading

This chapter presented the functional decomposition design technique, where every level of the design is decomposed into submodules, each of which is the domain of the next lower level. The inputs, outputs, and functionality must be determined for a given module. Appling the process in Figure 5.1 and following the guidelines in Section 5.3 should aid in the application of functional decomposition. Functional decomposition is applicable to a wide variety of systems, and in this chapter designs of analog electronics, digital electronics, and software were examined.

Nigel Cross presents a good overview of the functional decomposition method with application to mechanical systems [Cro00], but with less focus on the description of the functional requirements than presented here. The work by Stevens et al. [Ste99] is interesting reading that gives an understanding of the evolution of structured design. It delves into the concepts of coupling and cohesion. Coupling and cohesion are also addressed well in the book by Jalote [Jal97]. An in-depth treatment of structured systems design is found in <u>The Practical Guide to Structured Systems Design</u> [Pag88]. This guide also integrates data flow diagrams with functional techniques. Finally, the thermometer design example was inspired by Stadtmiller's book [Sta01] on electronics design.

5.11 Problems

5.1 Describe the differences between *bottom-up* and *top-down* design.

5.2 Develop a functional design for an audio graphic equalizer. A graphic equalizer decomposes an audio signal into component frequency bands, allows the user to apply amplification to each individual band, and recombines the component signals. The design can employ either analog or digital processing. Be sure to clearly identify the design levels, functional requirements, and theory of operation for the different levels in the architecture.

The system must

- Accept an audio input signal source, with a source resistance of 1000 Ω and a maximum input voltage of 1 V peak-to-peak.
- Have an adjustable volume control.
- Deliver a maximum of 40 W to an 8 Ω speaker.
- Have four frequency bands into which the audio is decomposed (you select the frequency ranges).
- Operate from standard wall outlet power, 120 V rms.

5.3 Develop a functional design for a system that measures and displays the speed of a bicycle. Be sure to clearly identify the design levels, functional requirements, and theory of operation for each level.

The system must

- Measure instantaneous velocities between zero and 75 miles per hour with an accuracy of 1% of full scale.

- Display the velocity digitally and include one digit beyond the decimal point.
- Operate with bicycle tires that have 19-, 24-, 26-, and 27-inch diameters.

5.4 Draw a structure chart for the following C++ program:

```
void IncBy5(int &a, int &b);
int  Multiply(int a, int b);
void Print(int a, int b);

main() {
    int x=y=z=0;
    IncBy5(x,y);
    z=Mult(x,y);
    Print(x,z);
}
void IncBy5(int &a, int &b) {
    a+=5;
    b+=5;
    Print(a,b);
}

int Multiply(int a, int b) {
    return (a*b);
}

void Print(int a, int b) {
    cout << a << ", " << b;
}
```

5.5 Develop a functional design for software that meets the following requirements.

The system must

- Read an array of floating point numbers from an ASCII file on disk.
- Compute the average, median, and standard deviation of the numbers.
- Store the average, median, and standard deviation values on disk.

The design should have multiple modules and include the following elements: (a) a structure chart, and (b) a functional description of each module.

5.6 Describe in your own words what is meant by coupling in design. Describe the advantages of both loosely and tightly coupled designs.

5.7 **Project Application.** Develop a functional design for your project. Follow the presentation guidelines in Section 5.9 for communicating the results of the design.

Chapter 6 System Design II: Behavior Models

Genius is 1% inspiration and 99% perspiration. —Thomas Edison

The functional decomposition technique examined in Chapter 5 is a powerful modeling tool for system design that is applicable for describing input, output, and transform behavior. However, that approach by itself is limited in its descriptive ability. This was apparent in the digital stopwatch example that required the use of state diagrams, in addition to functional decomposition, to fully articulate the design. A state diagram is an example of a model, a standardized abstraction of a system. Models allow systems to be described without having to determine all of the implementation details. All models are not the same—they come in a variety of forms and each serves a different intention.

This chapter provides an overview of other design tools for describing system behavior, with an emphasis on computing systems. The first tool examined is the state diagram. This is followed by the flowchart, which describes algorithmic processes and logical behavior. Two modeling languages for information and data handling—data flow diagrams and entity relationship diagrams—are then examined. The final topic is the Unified Modeling Language, which is a collection of system views for describing behavior.

Learning Objectives

By the end of this chapter, the reader should:
- Be familiar with the following modeling tools for describing system behavior: state diagrams, flowcharts, data flow diagrams, entity relationship diagrams, and the Unified Modeling Language.
- Understand the intention and expressive power of the different models.
- Understand the domains in which the models apply.
- Be able to conduct analysis and design with the models.
- Understand what model types to choose for a given design problem.

6.1 Models

From the previous chapter we know that the top-down design process starts with an abstraction of the system to be built. This initial design is called an abstraction because it captures the essential characteristics of the system without specifying the underlying physical realization. An abstraction that is expressed in a standardized and accepted language is called a model. In other words, a model is a standardized representation of a system, process, or object that captures its essential details without specifying the physical realization. A modeling language does not have to be formed from letters and words—often the words are graphical symbols. You are already familiar with many different models from everyday life such as blueprints, a diagram of a football play, knitting instructions, electrical schematics, and mathematical formulas to name a few. In order to be effective, a model should meet the following properties [Sat02]:

- *Be abstract.* This means that the model should be independent of final implementation and that there should be multiple ways of implementing the design based upon it.
- *Be unambiguous.* A model should have a single clear meaning in terms of describing the intended behavior.
- *Allow for innovation.* Models should encourage exploration of alternative system implementations and behaviors.
- *Be standardized.* Standardization provides a common language that can be understood by all. Designers should be wary of developing their own models that are ill-defined and not commonly understood.
- *Provide a means for communication.* A model should facilitate communication within the design team and with nontechnical stakeholders.
- *Be modifiable.* A model should make design modifications relatively easy.
- *Remove unnecessary details and emphasize important features.* The intent is to simplify the design for ease of understanding. The most highly detailed information is typically identified in the detailed design.
- *Break the overall problem into subproblems.* Most problems are too complex to be handled directly and must be decomposed into subsystems. This produces the design hierarchy.
- *Substitute a sequence of actions by a single action.* This allows understanding of the overall larger behavior, which can then be examined at other levels. This supports the ability to break a design into subproblems.
- *Assist in verification.* A model should aid in demonstrating that the design meets the engineering requirements.

- *Assist in validation.* Validation is the process of demonstrating that the needs of the user are being met and the right system is being designed. The model should facilitate discussion with all stakeholders to ensure it meets everyone's expectations.

In order to meet these properties, most models have an *object type,* which is capable of encapsulating the actual components used to construct the target system. In order to capture the dependence of objects on one another, models typically have a *relationship type*. Finally, models have an *intention,* which is the intended class of behavior that it describes. For example, the intention of a circuit schematic and the schematic of a football play are entirely different. Since models are built with different intentions, it is possible to choose the wrong model for a particular system—it would surprise a football team to see a play represented with resistors and capacitors!

Since models capture the essential details of a system in a standardized way, they are an ideal way to describe the functionality of a system at all levels of detail. In Chapter 5 the predominate method of describing functionality was with words. However, there are languages that describe system behavior. We start by examining state diagrams.

6.2 State Diagrams

State diagrams describe the behavior of systems with memory. A system with memory is able to modify its response to inputs based on the state of the system. The *state* of a system represents the net effect of all previous inputs to the system. Since the state characterizes the history of previous inputs, it is often synonymous with the word *memory*. Intuitively, a state corresponds to an operating mode of a system, and inputs are associated with transitions between states. To determine if a system has memory, ask the following question—"Can the same input produce different outputs?" If the answer to this question is yes, then the system has memory and can be modeled with a state diagram.

A state diagram is a drawing that consists of states and transition arcs, as shown in Figure 6.1. Each state is represented as a rectangle with rounded edges with the name of the state written inside. Whenever possible, states should be given meaningful names. When there exists the possibility for ambiguity in the names, a table should be created that identifies the state names and their associated meanings. There are special circle symbols for both initial and final states. Transitions are drawn as arrows from a source state to a destination state. Since inputs cause the transitions between states, the arrows are labeled with their associated inputs. The outputs are listed directly in the state since it is assumed that the outputs are associated with states.

Figure 6.1 Symbols used in state diagrams.

As an example, consider a vending machine that accepts nickels and dimes and dispenses a piece of candy when 25 cents has been deposited. This vending machine can be modeled by a state diagram because the response of the machine to a coin depends on how much money has been deposited so far.

In order to give a more complete description of the vending machine, the state diagram is embedded into the function table template introduced in Chapter 5, as shown in Figure 6.2. This table lists the inputs and outputs of the vending machine along with the behavior represented by a state diagram.

Module	Vending Machine Control Unit
Inputs	- Nickel: Signals that a nickel has been deposited. - Dime: Signals that a dime has been deposited.
Outputs	- Reset: Signals the FSM to return to the initial/reset state. - Vend: Signal to dispense candy.
Functionality	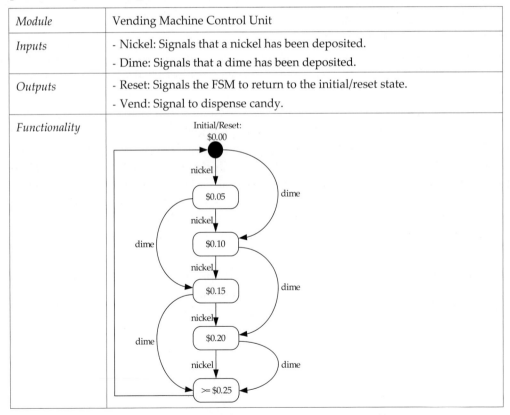

Figure 6.2 A state diagram for a simple vending machine.

The initialization/reset state is labeled $0.00, while the action of the machine (dispense or not dispense candy) is associated with the states—the only state that dispenses candy is $0.25. There are three types of transitions shown in this state diagram. The two labeled nickel and dime are associated with depositing those coins. The unlabeled transition from state $0.25 to state $0.00 is called an unconditional transition. A system with an unconditional transition between two states is assumed to remain in the first state for a defined time period before automatically moving to the second state. In this case it ensures that the machine dispenses a single candy before going on to the next transaction. It is common practice in state diagrams to assume that any unspecified input conditions cause the system to remain in the current state. For example, if no coin is inserted while the system is in state $0.20, the system remains in state $0.20. Finally, note that since this machine does not dispense change, it can overcharge a customer for candy. This would not be a popular vending machine with users!

6.3 Flowcharts

The intention of a *flowchart* is to visually describe a process or algorithm, including its steps and control. Flowcharts are often scoffed at as being old-fashioned and overly simple—these criticisms are actually strengths. Since flowcharts have been around for a long time, they are easily recognized and understood. Furthermore, being simple makes them accessible to a wide audience. Because of their simplicity, they are used in a great number of applications, including nontechnical ones such as the description of business processes.

Some of the primary symbols used in flowcharts are shown in Figure 6.3. The names of the starting and ending steps of a flowchart are represented by ovals known as terminators. Individual processing steps are written inside rectangles, while a process step that is elaborated by another flowchart is drawn as a rectangle with double sides. Elaborated processes allow the representation of hierarchy in the design. Certain points in a flowchart can lead to alternative destinations as repre,sented by a decision or conditional symbol (diamond). The condition that determines the next step is written inside the diamond and the possible values of the condition are written on the arcs leaving the conditional step. As shown in Figure 6.3 there are multiple ways to indicate data stores for retrieving or saving data.

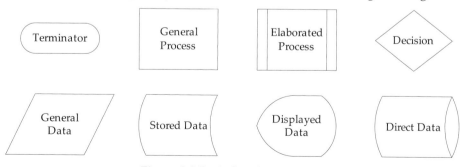

Figure 6.3 Basic flowchart symbols.

As an application, consider an embedded computer system that monitors the light level of its environment as described by the flowchart in Figure 6.4. The algorithmic process of the flowchart is easy and intuitive to understand. The system reads the current light value, stores it into an array, and then computes an average. In a complete system description, there would be a second flowchart describing how the system determines the average of the values of the light samples, since this is identified as an elaborated process. The system then waits 1 ms, writes the average to a terminal (display device), and then checks for a key-press. Light levels continue to be monitored in the absence of a key-press, otherwise the light monitoring process halts.

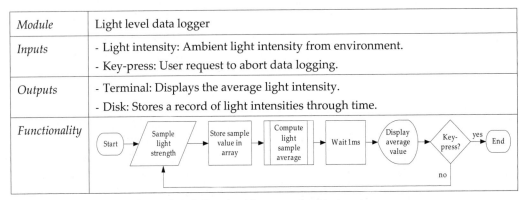

Figure 6.4 A flowchart for an embedded system.

Flowcharts are an intuitive way to describe algorithmic processes. Limiting the complexity of a flowchart to between 10 and 20 steps enables the sequence of actions to be quickly comprehended while eliminating unnecessary details. Flowcharts are not able to represent the structure of data being manipulated and they are not particularly good at representing concurrent processes. For this we need data flow diagrams.

6.4 Data Flow Diagrams

The intention of a *data flow diagram* (DFD) is to model the processing and flow of data inside a system. It is a function-oriented approach that is similar to functional decomposition—the processes inside a DFD accept data inputs, transform them in some way, and produce output data. A DFD is often used for the analysis of information systems due to its data emphasis, but can be broadly applied to electrical and computer systems. It differs from functional decomposition in that functional decomposition is often closer to the implementation of the design, whereas the DFD models the system from a data point of view. A DFD is fundamentally different from a flowchart in that it does not encapsulate control and sequencing information, but allows multiple processes running concurrently. There are four symbols, shown in Figure 6.5, that are used in a DFD:

1) *Processes.* A rectangle with rounded corners that describes a useful task or function. They perform a transformation on the data.
2) *Data flows.* An arrow representing a data relationship between two processes.
3) *Data stores.* An open rectangle representing a data repository.
4) *Interfaces.* A square describing external agents or entities that use the system. They are also referred to as sources and sinks.

Figure 6.5 Data flow diagram symbols.

Like the general design process, DFDs are successively refined from the top down. That is, there is a single top level (or Level 0) DFD describing the entire system and the interfaces and data stores that it interacts with. The rules for constructing a DFD are fairly intuitive. A process must have at least one input and one output. The refinement of a process at level N must have the same inputs and outputs as the process at level $N-1$. Data must be transformed in some way by a process. This process of refinement continues until a satisfactory level of detail is reached.

An example of a Level 1 DFD for a video browsing system is shown in Figure 6.6. Video databases are typically very large; because of their size, it is usually cumbersome to preview videos and extract important information. A solution to this problem is to apply image analysis techniques to identify shots (continuous scenes without a break) in a video and store both the location of the shot boundaries in the video and key frames that summarize each shot. The collection of key frames is known as a storyboard. The storyboards are stored in an annotation database that is much smaller in size than the original video database. Typically the user of a video browsing system has the ability to preview the storyboard and select shots from it to view. This example shows the data flow for such a system.

The data processing is readily apparent from the DFD. Videos in the database are processed to extract shot boundaries and key frames, which are stored in the annotation database. The user can submit requests to view a storyboard, which is retrieved from the annotation database. When a shot is selected from the storyboard, the user can preview the original shot, which is retrieved from the video database.

118 Design for Electrical and Computer Engineers

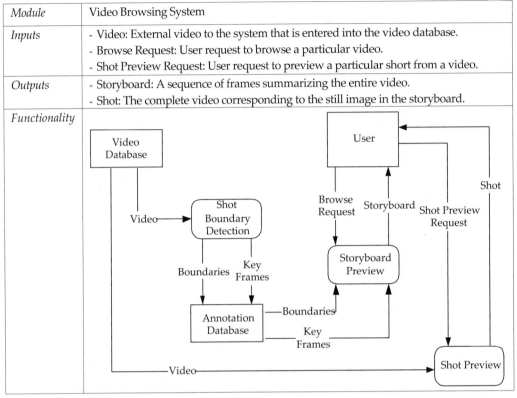

Figure 6.6 Level 1 data flow diagram for a video browsing system.

There are a few important points to note about DFDs. They are solution independent, specifying the behavior on a data flow basis. Specific information on the data flows is defined in a formal language known as a *data dictionary* (it is not covered here). There can be concurrent processes represented in a DFD. In the video browsing example, multiple people can use the system simultaneously, and there is no implied sequencing between when the shot boundary detection process is run and the storyboards are previewed. It is clear that the video must undergo shot detection before its storyboard can be viewed. This information is listed in something known as an *event table*. The event table for this example is shown in Table 6.1.

Table 6.1 Event table for the video browsing system.

Event	Trigger	Process	Source
Annotate Video	New Video Arrival	Shot Boundary Detection	System
View Storyboard	Browse Request	Storyboard Preview	User
View Shot	Shot Preview Request	Shot Preview	User

An *event* is an occurrence at a specific time and place that needs to be remembered. Events can be classified into temporal, external, and state. A *temporal event* is one that happens because the system has reached some critical time. In this example, the generation of shot boundaries could occur for all new videos in the database at a specified time each day, but in this case it occurs whenever a new video is added to the database. An *external event* occurs outside the system boundary by a system user, in this case requesting either a storyboard or a video preview. A *state event* is the result of something changing within the system. Associated with each event is a *trigger*, the cause of the event. Each event has a process that is associated with it. Finally, associated with each event is a *source*, the entity responsible for triggering it.

6.5 Entity Relationship Diagrams

A database is a system that stores and retrieves data, and it is modeled by an *entity relationship diagram* (ERD). The intention of an ERD is to catalog a set of related objects (entities), their attributes, and the relationships between them. The entities and their relationships are real, distinct things that have characteristics that need to be captured. The design of a database starts by describing the entities, their attributes, and the relationship between entities in an ERD. In order to ask meaningful questions about the data, the entities need to be related to one another. For example, a list of students and a list of courses by themselves have limited utility. However, by introducing a relationship between these two entities we can ask questions such as "How many students are taking the microelectronics course?" The three elements used in the ERD modeling language are:

1) *Entities.* They are generally in the form of tangible objects, roles played, organizational units, devices, and locations. An instance is the manifestation of a particular entity. For example, an entity could be Student while an instance would be Kristen.
2) *Relationships.* They are descriptors for the relationships between entities.
3) *Attributes.* They are features that are used to differentiate between instances of the entities.

Let's consider an ERD describing academic scheduling at a college. The process typically starts by interviewing the end users and identifying the entities and their attributes. Assume that the result of this process is that the college wants to store data about three entities: Students, Courses, and Departments. The process of building an ERD starts by determining the relationships between entities. One way to do this is to build an *entity relationship matrix* as shown in Table 6.2. The entities constitute both the row and column headings, and the matrix entries represent the relationship, if any, that exists between entities. This is similar to the pairwise comparison matrix in Chapter 2, where user needs were systematically compared. Relationships are bidirectional because they have two participating entities.

Table 6.2 Entity relationship matrix.

	Student	Course	Department
Student		takes many	majors in one
Course	has many	can require many / can be the prerequisite for many	is offered by one
Department	enrolls many	offers many	

From Table 6.2 we can see that a Student can take many courses and a Course has many students in the Student-Course relationship. A *cardinality ratio*, associated with each relationship, describes the multiplicity of the entities in a relationship. For example, the Student-Course relationship is many-to-many, or M:M, because one student can take *many* courses and one course is taken by *many* students. The relationship between Student and Department is M:1 since a Department enrolls *many* students, but a student can major in only *one* Department. A recursive relationship, prerequisite, exists between the Course entity and itself because a course may have *many* other courses as prerequisites and may be the prerequisite for *many* other courses. Thus the prerequisite relationship has an M:M cardinality. It needs to be noted that in this example the needs of the college required relationships between all pairs of entities. If, for example, the college did not need to keep track of the students' majors, then the relationship between Student and Department would be left blank in the entity relationship matrix.

The resulting ERD is shown in Figure 6.7. The relationships are identified by the diamond-shaped symbols; the entities are denoted by the rectangles with their name at the top. The cardinality ratio is labeled on the links between the relationships and entities. Another piece of information shown in the ERD is the attributes associated with each entity. An attribute is a feature or characteristic of an entity that needs to be remembered. There are many different types of attributes; however, the most important are ***key attributes*** (which are underlined) which uniquely identify instances For example, the identification number (ID) attribute of a Student is a key attribute.

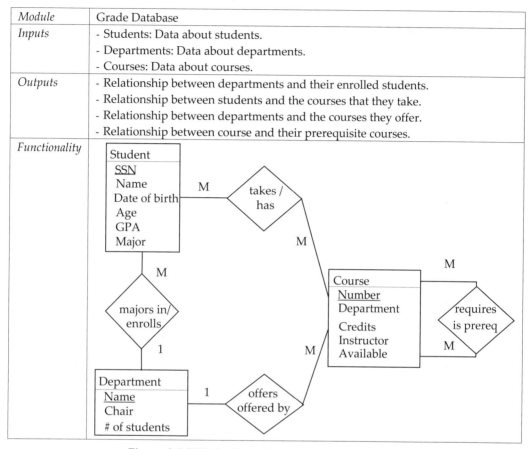

Figure 6.7 ERD for the college database system.

The ERD allows for an easy interpretation of the relationships between entities as well as their attributes. Although beyond the scope of the discussion here, the ERD is a formal language that can be used to automatically generate the database structure. The relationships in the ERD allow queries to be asked of the resulting database. For example, the college could derive a student's course schedule from the database from the relationship between Student and Course.

6.6 The Unified Modeling Language

The *Unified Modeling Language* (UML) was created to capture the best practices of the object-oriented software development process. However, it has valuable modeling tools that can be applied more generally to many types of systems. The objective of this section is to provide an overview of UML and the six different system views that it encompasses. A caveat for the reader—UML is complex, and complete coverage is beyond the scope of this book. In order to fully understand its subtleties and nuances, the reader is advised to consult the references provided in Section 6.8.

In order to aid in the explanation of the different views we will create a UML model of a system called the Virtual Grocer, or v-Grocer for short. The v-Grocer enables a user to order groceries from home and have them delivered. The concept is to provide users with a bar code scanner that is connected to their home computer along with application software. When users run out of an item, they scan in the Universal Product Code (UPC). When users want to order groceries, they connect to the Internet, log into the grocery store web server, enter the quantity for the scanned items, and place their order. Once the order is completed, they are billed and the groceries are delivered to their houses at a prearranged time.

6.6.1 Static View

Object-oriented design (OOD) is fundamentally different from functional software design in that it emphasizes objects instead of functions. **Objects** represent both data (attributes) and the methods (functions) that can act upon the data. An object represents a particular instance of something known as a *class*, which defines the attributes and methods. An object encapsulates all of this information. The data and methods that an object encapsulates are available only to that object by default. Thus, other objects cannot change that state of an object unless given specific permission to do so. This improves the reliability and maintainability of software systems, because changes made to the internal representation of a class are not seen by methods outside of the class.

The intention of the *static view* is to show the classes in a system and their relationships. The static view is characterized by a *class diagram*. A very simple class diagram with a single class is shown in Figure 6.8. The specification of a class has three parts: a name, a list of attributes, and a set of methods. The name of the class for this example is Customer and it has three attributes: Name, Address, and CustId. It has one method associated with it, ChangeAddr(), which can change the Address attribute. A class, just like an entity in an ERD, is a generalization of a set of a particular thing or instance. For example, the Customer class could have a particular instance called Ms. Robinson.

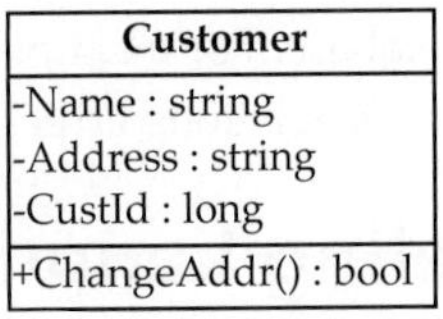

Customer
-Name : string
-Address : string
-CustId : long
+ChangeAddr() : bool

Figure 6.8 Class diagram notation. The class has a label (Customer), attributes (Name, Address, and CustId), and methods (ChangeAddr()).

Classes are related to one another by relationships that define how classes interact with each other. For example, if a class is a subset of another class (a kind of relation) then the subclass inherits all the attributes and methods of the *superclass*. In UML, there are a host of relationships; among these are generalization, composition, and associations. Two classes are related by a *generalization relationship* when one is a subset of the other. Two classes are related by a *composition relationship* when one is composed of members of the other. Two classes are related by an *association relationship* whenever they need to send messages to one another. Relationships are drawn as lines connecting the two participant classes. The terminals of the line have different shapes depending on the type of relationship. Just like an ERD, the relationships between classes have cardinality defined by the rule base. In order to illustrate these points, Figure 6.9 shows the class diagram for a portion of the v-Grocer system.

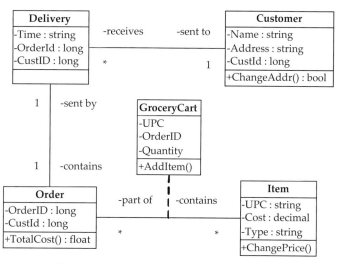

Figure 6.9 Class diagram for the v-Grocer.

The class diagram has five classes—Customer, Delivery, Order, Item, and GroceryCart. The GroceryCart class is derived from the many-to-many relationship that exists between the Order and Item classes. Just like the rules for an ERD, the cardinality of a relationship is read at the end of the association in which it is involved. The only difference is that the many cardinality in a class diagram is denoted by a * symbol. For example, a delivery is sent to one customer, while a customer may receive many deliveries.

6.6.2 Use-Case View

The intention of the *use-case view* is to capture the overall behavior of the system from the user's point of view and to describe cases in which the system will be used. It is characterized by a *use-case diagram*, and an example for the v-Grocer is shown in Figure 6.10. There are only two symbols employed in a use-case diagram, actors and use-cases. Actors, drawn as stick figure people, are idealized people or systems that interact with the system. For this example the actors are customers, delivery people, the database, clerks, and the web server. Use-cases are drawn as ovals in the diagram, and in this example, they are WebOrder, DeliverOrder, and `AssembleOrder`. A use-case is a particular situation when actors use the system and is usually represented by a sequence of actions that will be performed by the system. This sequence of actions typically represents a high level of functionality. Every actor that interacts with a particular use-case is connected to it with a line. Finally, the entire collection of use-cases is enclosed in a rectangle with the actors outside. This rectangle represents the system boundary and consequently is labeled with the name of the system.

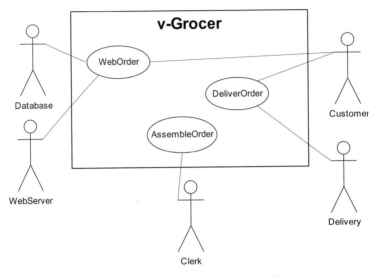

Figure 6.10 Use-case diagram for the v-Grocer.

From Figure 6.10 it is apparent that a `Customer`, the `WebServer`, and `Database` interact when creating a `WebOrder`. Use-cases are often described in a table as shown for this particular example in Table 6.3.

Table 6.3 `WebOrder` use-case description.

Use-Case	`WebOrder`
Actors	`Customer`, `Database`, and `WebServer`
Description	This use-case occurs when a customer submits an order via the `WebServer`. If it is a new customer, the `WebServer` prompts them to establish an account and their customer information is stored in the `Database` as a new entry. If they are an existing customer, they have the opportunity to update their personal information.
Stimulus	Customer order via the `GroceryCart`.
Response	Verify payment, availability of order items, and if successful trigger the `AssembleOrder` use-case.

Use-cases focus on a very high level of functionality and describe who will interact with the system and how they will interact. Given their high level of abstraction, they can easily be incorporated into a variety of electrical and computer engineering projects. They are also simple and easy to understand and thus can be incorporated into presentations to nontechnical audiences. Finally, use-cases are simply not busy work—they are fundamental to the development of other UML models.

6.6.3 State Machine View

The *state machine view* is characterized by a *state diagram*, as was discussed in Section 6.2. Again, the intention of a state diagram is to describe systems with memory. Figure 6.11 shows the state view of a customer logging into the v-Grocer web server.

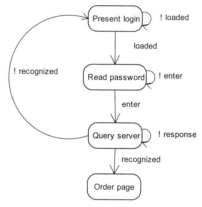

Figure 6.11 State diagram for the v-Grocer customer login.

6.6.4 Activity View

The intention of the *activity view*, characterized by an *activity diagram*, is to describe the sequencing of processes required to complete a task. In UML, the tasks elaborated are the individual use-cases identified in a use-case diagram. Since more than one actor may be involved in completing a task, an activity diagram can express the concurrent nature of tasks. An activity diagram is composed of states, transitions, forks, and joins. Figure 6.12 contains an activity diagram for the v-Grocer order delivery system. The diagram gives a clear visual picture for the activities that need to be completed for an order to be packed and delivered. After a complete order is placed, the flow is forked into two concurrent processing branches. One of these branches is for completion of the order, while the other addresses its scheduling, delivery, and coordination with other deliveries.

When both of those processes are completed, they are joined together and then the delivery run is made.

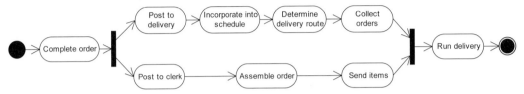

Figure 6.12 Activity diagram for order processing and delivery for the v-Grocer.

6.6.5 Interaction View

The intention of the ***interaction view*** is to show the interaction between objects. It is characterized by collaboration and sequence diagrams. In an object-oriented system, tasks are completed by passing messages between objects. The interaction view shows how messages are exchanged in order to accomplish a task. These tasks are usually the use-cases. Since messages are sent through time, the interaction view must be able to express the concept of order. We start by examining collaboration diagrams.

Two objects collaborate together in order to produce some meaningful result. A *collaboration diagram* shows the sequencing of messages that are exchanged between classes in order to complete a task. It consists of the classes that participate in the realization of the task and the messages exchanged. The messages are drawn as arrows from the sending object to the receiving object. The message arrows are labeled with the name of the message and its numerical position in the sequence of messages exchanged to realize the task. For example, Figure 6.13 shows how the `WebOrder` use-case is realized using the `Customer`, `WebServer`, and `Database` classes. Note, `WebServer` and `Database` are introduced as new classes and are not shown as part of the class diagram in Figure 6.9.

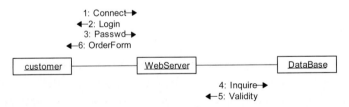

Figure 6.13 Collaboration diagram for the `WebOrder` use-case.

The collaboration diagram is similar in form to the class diagram, the difference being that the relationships are annotated with the messages that are exchanged. Consequently, the collaboration diagram aides the developers in understanding and implementing the methods used between classes. In order to emphasize the order in which messages are exchanged, a developer can also use a sequence diagram.

A *sequence diagram* contains the same information as the corresponding collaboration diagram. Where the collaboration diagram emphasizes the objects that interact to produce a behavior, a sequence diagram emphasizes the message order that produces a behavior. As shown in Figure 6.14, the classes that participate in the behavior are listed in a row at the top of the diagram. From each class a dotted vertical line is drawn downward that represents the lifeline of its class (the vertical axis represents time). When an object is actively requesting or waiting for information from another object, a rectangle is drawn over the dotted lifeline of the object. The message is drawn as an arrow from the sending object to the receiving object and labeled with the name of the message. The activity diagrams can be applied generally to electrical and computer systems to show the interaction between entities, particularly the sequencing of messages.

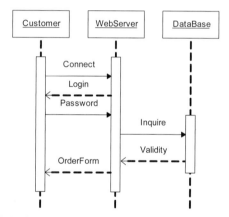

Figure 6.14 Sequence diagram for the `WebOrder` use-case.

6.6.6 Physical View

The intention of the *physical view* is to demonstrate the physical components of the system and how the logical views map to them. The physical view is characterized by a component and deployment diagram. A *component diagram* shows the software files and the interrelationships that make up the system. Software files are shown in rectangles. Lines connect together files that need to communicate. A *deployment diagram* shows the hardware and communications components that will be used to realize the system. The hardware components are drawn as cubes and labeled with their names. Figure 6.15 shows a combined component and deployment diagram for the v-Grocer system. The software files are the `Browser`, `v-Grocer`, `Apache`, and `Oracle`, while the hardware components are the `Customer`, `WebServer`, and `Database`.

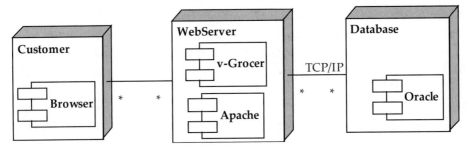

Figure 6.15 The combined component and deployment diagram for the v-Grocer. The software files are the `Browser`, `v-Grocer`, `Apache`, and `Oracle`, while the hardware components are the `Customer`, `WebServer`, and `Database`.

6.7 Project Application: Selecting Models

Chapters 5 and 6 have presented a variety of models for representing the behavior of systems. The design team needs to select the correct combination of tools to properly describe a design for its eventual implementation. Table 6.4 provides a summary of the models and their different intentions.

Table 6.4 Guidance for model selection.

Model	Intention
Functional Decomposition	To describe the input, output, and functional transformations applied to information (electrical signals, bits, energy, etc.) in a system. It is broad in application. This often provides a view that is close to the actual system implementation. See Chapter 5.
State Diagram	To describe the behavior of systems with memory. It is very flexible when it comes to application. It is often applied to digital design, but state diagrams can also be used to describe the high-level behavior of complex systems. The only prerequisite on their use is that the system has memory. See Section 6.2.
Flowchart	To describe a process or algorithm, including its steps and control. It is applicable to many problem domains from software to describing business practices. See Section 6.3.

Data Flow Diagram	To model the processing, transformation, and flow of data inside a system. It is typically supplemented by an event table describing all possible events and the resulting actions. Creating a DFD requires the designer to carefully think about the uses of the system and how the system is to react to external users and events. See Section 6.4.
Entity Relationship Diagram	To catalog a set of related objects (entities), their attributes, and the relationships between them. ERDs capture the entities and relationships of a portion of the world into a formal data model. The graphical language describes the attributes of the entities and the cardinality of the relationships. ERDs are formal enough to be unambiguously translated into a complete description of a database system. See Section 6.5.
The Unified Modeling Language. The intention of UML is to describe complex software systems. However, certain views are well suited to describing systems at a high level and are applicable to many domains. The process of viewing a system from the six perspectives listed below decreases the chances that crucial details of the design will be overlooked.	
Class Diagram	To describe classes and their relationship in an object-oriented software system. Class diagrams are primarily for software design and are not easily accessible to a nontechnical audience. See Section 6.6.1.
Use-Case Diagram	To capture the overall behavior of the system from the user's point of view and describe cases in which the system will be used. See Section 6.6.2.
State Machine	This is essentially the same as the state diagram, but is also a formal UML view. See Sections 6.2 and 6.6.3.
Activity Diagram	To describe the sequencing of processes required to complete a task. Composed of states, transitions, forks, and joins. Can show concurrent processes. See Section 6.6.4.
Interaction Diagram	To show the interaction and passing of messages between entities in a system. Characterized by both collaboration and sequence diagrams. See Section 6.6.5.
Physical Diagram	To describe the arrangement and connections of the physical components that constitute a system. In the case of UML, it is characterized by a component and deployment diagram. See Section 6.6.6.

6.8 Summary and Further Reading

Models provide a convenient method of describing a system at a high level of abstraction. This allows a system to be described without having to determine all of the implementation details. All models are not the same—they come in a variety of forms and each is designed to serve a different intention. The properties of effective models were presented, and we saw that they are similar to those of an engineering requirement. Models should encourage innovation by allowing the exploration of alternative implementations. Since models are built from abstractions of the actual system components, they are an effective means for communicating with nontechnical stakeholders. Finally, models provide an excellent means of documenting the development of a design from the highest level down to the detailed design.

This chapter presented a variety of models for describing system behavior that included state diagrams, flowcharts, data flow diagrams, entity relationship diagrams, and the Unified Modeling Language. Table 6.4 provides a quick reference that describes the intention and application of the models examined in both Chapters 5 and 6.

Flowcharts have been around for quite some time, and an early work describing them is <u>Flowcharts</u> by Chapin [Cha71]. The book <u>Programming Logic and Design</u> [Far02] covers flowcharts extensively and their application in programming. State diagrams are fundamental to the electrical and computer engineering field, and further information on them is available in virtually any introductory digital design textbook. Data flow diagrams and entity relationship diagrams are common tools used in information systems analysis, and design and can be further explored in <u>Systems Analysis and Design</u> [Sat02] and <u>Fundamentals of Database Systems</u> [Elm94]. Two references for UML are <u>The Unified Modeling Language Reference Manual</u> [Rum98] and <u>Schaum's Outline of UML</u> [Ben01].

6.9 Problems

6.1 Why is it important for a model to separate the design of a system from its realization?

6.2 Classify each of the following as either a model, not a model, or sometimes a model. Justify your answer on the basis of the definition and properties of a model.
   a) A diagram of a subway system
   b) A computer program
   c) A football play
   d) A driver's license
   e) A floor plan of the local shopping mall (a "you are here" diagram)
   f) An equation
   g) A scratch 'n' sniff perfume advertisement in a fashion magazine
   h) The 1812 Overture
   i) A braille sign reading "second floor"
   j) Sheet music for the Brandenburg Concerto
   k) The United States Constitution
   l) A set of car keys
   m) The ASCII encoding of an email message

6.3 Which of the following systems has memory? Justify your answer using the concepts of input, output, and state.
   a) An ink pen

b) A resistor
c) A capacitor
d) A motorized garage door
e) An analog wristwatch
f) The air pressure in an air compressor
g) The thermostat that controls the furnace in a house
h) A light switch
i) A political system
j) The temperature of a large lake
k) A book
l) A computer's hard drive

6.4 A can of soda has memory. Your objective is to figure out what characteristic of the can is the state variable and what input causes it to change. Using this information, draw a state diagram for a can of soda. Label the transition arcs with the input responsible for the transition. Hint: no special equipment is needed to elicit the change of state.

6.5 Consider the state diagram for the vending machine shown in Figure 6.2. Now assume that the system accepts nickels, dimes, and quarters. Also assume that it is capable of returning change to the user after a purchase. Create a state diagram that represents this new system. Make sure to define the output signals and their value for each state.

6.6 Use a state diagram to describe the high-level operation of the ChipMunk Recorder (CMR). The CMR records sounds and then plays them back at a variety of speeds, making a recorded voice sound like a high-pitched chipmunk. The CMR receives user input from a keyboard and an audio source. The behavior of the system is described as follows:

- When powered up, the CMR enters a wait state.
- If R is pressed the recorder begins recording.
- Any keypress will put the CMR back into the wait state.
- If S is pressed, the CMR is ready to change the playback speed. A subsequent numerical input between 1 and 5 will cause the playback speed to be changed to that value.
- Pressing the R key when in the adjust playback speed mode will cause the CMR to go to the wait state.
- Pressing a P key will cause the CMR to play back the recorded sounds. When done playing the entire recording, the CMR will loop back and start playing at the beginning.

- Any keypress while it is in the playback mode will cause the CMR to go back into the wait state.

Draw a state diagram describing the behavior of the CMR. Create a table that lists every state and its associated output.

6.7 Build a state diagram to describe the state of the tape cartridges used to back up a company's network drives. When new `unformatted` cartridges are received, they are immediately labeled with a unique ID. Before a tape is used, it is formatted and thereby turned into a `blank` tape. On the first day of the week a complete backup is made of the network drives, transforming blank tapes into `active` tapes. The active tapes, made every fourth week, are moved off site, making them `archival` tapes. Active tapes older than 3 months are assumed to have out-of-date information and are reformatted into blank tapes. Archival tapes more than 2 years old are reformatted and put back into circulation.

6.8 Build a flowchart to describe the operation of a microcontroller- (MCU-) based temperature regulating system. The system monitors the temperature of a heated environment, using thermistors, and regulates the temperature by turning fans on to cool the environment. The MCU periodically reads the temperature from each of the 64 different thermistors (each is driven by its own constant-current source) by selecting each through an analog multiplexor. The voltage is converted into an 8-bit digital value by the MCU's analog-to-digital converter. If any of the 64 thermistors exceeds a high-temperature threshold, the MCU uses a complex algorithm to determine the number of fans to turn on, otherwise all the fans are turned off.

6.9 Write an algorithmic description for each of the flowcharts below using `while`, `if`, or `do` statements.

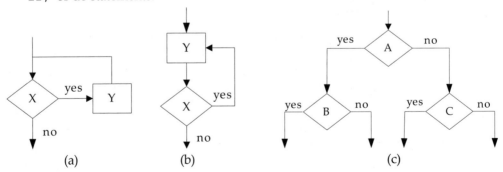

6.10 Create a flowchart that outlines how to crochet a two-tone blanket with a diagonal stripe across it as shown below. A blanket is crocheted by linking together a sequence of basic stitches. For the purposes of the flowchart assume that a basic stitch is an elaborate process. Basic stitches are made from either dark or light yarn. The blanket should be 100 stitches wide by 150 stitches high. The diagonal stripe runs at a 45 degree angle from the horizontal.

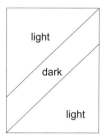

6.11 Build a data flow diagram and event table to represent an image archiving system for an art museum. The art museum maintains a database of digital images of paintings from museums all over the world. The following information is known about the image database system:

- Images are shared among museums in a participating network. Whenever a participating museum posts a new image, it sends a broadcast email to all participating museums in the network with the image attached as an email. It provides the name of the painting and artist in the body of the email. All new images received are added directly to the museum's own image database.

- When inserted into the museum's database, each image is provided with a tag identifying the name of the painting and the artist. Furthermore, this triggers an image analysis routine that classifies the image into a predefined category such as portrait, natural scene, and modern. Furthermore, it stores key features that are extracted from the image.

- The key features are used to identify and retrieve visually similar images from the museum's database. Another image processing algorithm is run that compares the visual similarity of the new image to all images in the database. This process produces a matching score of 0 to 100, which is stored.

- The museum's image database is available to visitors via computer kiosks placed throughout the museum. Kiosk users can retrieve and view images in one of three ways. First, they can specify the name of the artist or painting. Second, they can retrieve a class of images, such as modern. Third, once they have received a painting, they can submit a request to view visually similar images. The visually similar images are retrieved for viewing on the basis of matching score.

6.12 Build an ERD to keep track of the bicycle frames manufactured at a local company. The following are notes from an interview with the owner.

We custom-build bike frames to the dimension of each individual customer. When a customer comes in we take measurements of height, leg length, arm length, torso length, weight, and waist measurement. Since we have high customer satisfaction, our customers order new frames every several years. Hence we would like to date these measurements in order to track how a customer's body changes through time. Each frame is built on one set of measurements. Clearly, we need to keep track of our customer's contact information like name, address, phone number, and email address. We would like to know which employee built each frame. We would like to store basic information like name, address, phone, and SSN for each employee. Each frame is built by one employee using a variety of different titanium tubing. We have strict inventory control on all of our tubing and need to keep track of its grade, lot number, outer diameter (OD), inner diameter (ID), and manufacturer. Tubing is uniquely identified by its lot number. Finally we need to keep information on the frame. Each frame is given a unique serial number, and has a color, type, and dimensions.

6.13 Extend Problem 6.7 to create an ERD that captures data about the tape cartridges used in the backup system. Every Sunday night a full backup is made of all network drives. A full backup creates an identical copy of the network drives on the tape cartridges. Because of the large amount of information, a full backup requires many tapes. On the other nights of the week an incremental backup is made. An incremental backup stores only files modified since the last backup (either full or incremental). Incremental backups are much smaller than a full backup, and consequently many incremental backups fit on a single tape. A tape contains only full or incremental backup information—the unused portion of the last tape used for a full backup is never used to store incremental backups. Your company wants to keep track of tapes, full backups, and incremental backups. An ID and state should be tracked for each tape. For full backups, the system needs to track the creation date and the number of tapes used. For an incremental backup it should track the date it was made. The relationships between the backup type and the tape will capture which tapes participated in which backup. (Hint: the state of a tape should be an attribute of the tape entity—unformatted, blank, etc. They are not attributes and are possible values for the state attribute.)

6.14 **Project Application.** Develop behavior models that are applicable for describing your system design. Table 6.4 is provided to help in making the determination as to which models are applicable.

Chapter 7 Testing

A stitch in time saves nine. —Anonymous

Most systems undergo testing throughout their development and before they are delivered to the customer. Clearly, systems should be tested to ensure that they meet the engineering requirements. In fact, one of the desirable properties of an engineering requirement is that it be verifiable, or in other words, testable. The philosophy of testing is embodied in the quote above, which means that it is better to correct errors early, rather than wait until they become much larger problems later. As we saw in Chapter 1, the cost to correct problems increases exponentially with the lifetime of the project. Thus, testing should be considered throughout system development.

Testing means different things to different people. A field service technician, assembly line worker, and designer will have their own definitions and requirements from a test. In this chapter testing is examined from the perspective of a systems designer intent on checking that the system meets the engineering requirements. Along the way fundamental testing concepts like controllability and observability are explored. Approaches to debugging systems are provided, followed by templates for building unit tests, integration tests, and acceptance tests.

Learning Objectives

By the end of this chapter, the reader should:
- Understand the concepts of black box tests, white box tests, observability, and controllability.
- Understand the principles of debugging.
- Understand when a unit test is used and how it is constructed.
- Understand when an integration test is used and how it is constructed.
- Understand when an acceptance test is used and how it is constructed.

7.1 Testing Principles

The design process is really a continual increase in specificity from engineering requirements to the detailed design. We now consider the question of how to test that the resulting system meets the design requirements. One answer is based on a common testing model, the "test vee," shown in Figure 7.1. This model starts with the engineering requirements, proceeds to the implementation, and then onto testing. It emphasizes that every level of design has a corresponding level of test. What is not so clear from this model is that the testing process is actually split between the two halves of the test vee. Students typically think of testing as being exclusively confined to the right half of the test vee—build it then you test it. However, each test performed in the right half of the test vee must be carefully engineered during the development of the system in the left side of the test vee. An acceptance test plan should be written with the requirements specification, integration tests defined and written during the system design, and so forth.

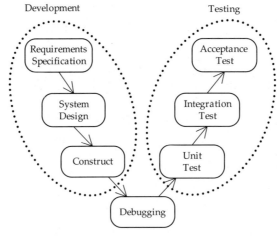

Figure 7.1 The test vee. Design stages are on the left and corresponding tests are on the right.

In our enthusiasm to complete a project many of us all too often rely on a "smoke test"—turn on a system to see if it works. The name of this test is a reference to what may happen to the system if the test fails—it burns up and smokes. Beyond being a potentially expensive way to test a system, a smoke test is not a systematic approach to verify that the system behaves as expected. Customers will not be impressed with "Hey, it didn't catch on fire!" as the test result. Clear tests need to be developed because:

- The test cases define exactly what the module must do.
- Testing prevents "feature creep," since the development of a module is complete when its test is passed.
- Test cases motivate developers by providing immediate feedback.
- Test cases force designers to think about extreme cases.
- Test cases are a form of documentation.
- Test cases force the designer to consider the design of the module before building it.

The test suite and its accompanying documentation contain important information about the behavior and organization of a system and its module. This gives tests a value beyond a role in showing that the system and its modules do not fail the tested conditions. The individual test cases show other engineers how to properly interface to a module, making that module more reusable. In addition, test documents can be used by other individuals in the organization such as technical writers, maintenance technicians, and technical trainers.

How should testing be done? Given enough time, a system could be tested by simply enumerating every conceivable input and observing the outputs. While in some cases this might be possible, in general it would take an unreasonable amount of time to perform such a test. Instead, tests are crafted in order to maximize the likelihood of finding errors.

7.1.1 Types of Testing, Observability, and Controllability

Tests fall into two general types—black box and white box tests. **Black box tests** are those that are performed without any knowledge of the system's internal organization. In a black box test, the testing is typically conducted by changing the inputs and comparing the system outputs to their expected values. The input and output values can be classified as typical, boundary, extreme, and invalid. These categories are illustrated by considering a system that converts Celsius temperatures to Fahrenheit. Typical inputs are values experienced during normal operation, say room temperature. Boundary values are encountered whenever the input or output changes in some significant way. For example, 0°C and −33.3°C mark the transition between positive to negative temperatures in Celsius and Fahrenheit respectively. Absolute zero represents an extreme value, because things can't get any colder. While these tests could be accomplished by enumerating every possible input to the system and observing the output, this would take an unreasonable amount of time. Hence, the test writer must elect candidate inputs to represent the behavior of the system over a range of possible inputs. An important goal of the test writer is to minimize the number of these equivalence classes while maximizing coverage of the input domain. Without a clear understanding of the internal organization of the system this is a challenging goal.

White box tests are conducted with knowledge of the internal working of the system. The idea of white box testing is to build tests which target specific internal nodes of the system to check that they are operating as expected. The tests should be written to check that node can handle typical, boundary, extreme and illegal situations.

One of the many goals in designing a system is to increase its testability. A design is ***testable*** when a failure of a component or subsystem can be quickly located. A testable design is easier to debug, manufacture, and service in the field. One way to increase the testability of a system is to increase controllability and observability. ***Controllability*** is the ability to set any node of the system to a prescribed value. ***Observability*** is the ability to observe any node of a system. In black box testing, both controllability and observability are low. In white box testing, controllability and observability may be higher, depending on the design.

Let's examine this further via the example of a simple transistor amplifier shown in Figure 7.2. The purpose of this circuit, known as the common-emitter amplifier, is to amplify the input signal, v_i, to produce a linearly proportional output signal, $v_o = A \times v_i$. The rectangular boundary in the figure represents a black box view of the system. In this view, the system power, V_{cc}, and ground would be applied to activate the circuit. The black box testing would consist of checking supply and ground voltages, varying the input signal, and observing the output signal. Again, this is a low-controllability and low-observability situation.

White box testing utilizes knowledge of the internal workings of the design. In designing a transistor amplifier, there are two major points to consider—the DC bias voltages in the circuit and its AC, or time-varying, amplification behavior. The two behaviors are related since the AC behavior depends upon proper DC biasing of the circuit. During detailed circuit design, the expected DC voltages for different nodes in the circuit would be determined. Thus, a white box test would consist of first checking the power supply and ground voltages as was done in the black box case. The next step would be different in that the node voltages (V_B, V_C, V_E) would be checked to see if they meet the expected design values. This indicates a high degree of observability. However, the controllability is not significantly better than in the black box case. This is because the internal DC node voltages in the circuit cannot arbitrarily be changed without negatively changing the operation of the circuit.

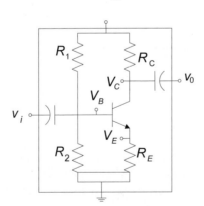

Figure 7.2 Transistor amplifier design.

7.1.2 Stubs

A *stub* is a device that is used to simulate a subcomponent of a system. This might be done for either of two reasons: the subcomponent has not yet been built or the risk of damaging the subcomponent warrants using a stand-in. Typically, stubs are used to simulate inputs or monitor outputs of the **unit under test** (UUT). Both hardware and software stubs can be used in designing a system. In software testing, stub routines are developed to either call other functions or act as those to be called by the unit under test.

Consider a hardware example, the transistor amplifier in Figure 7.2. Assume that the circuit is ultimately to be integrated into a larger system. The input to this system is a time-varying source with certain resistive and capacitive characteristics, while the output is connected to another system with a known input resistance range. The stubs used for testing in this system are shown in Figure 7.3. On the input side is a function generator, an off-the-shelf component, connected to a resistor and capacitor that models the expected characteristics of the final system. The stub on the output side is simply a resistor, whose value can be varied over the expected load.

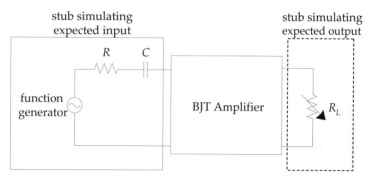

Figure 7.3 The use of stubs for testing a transistor amplifier circuit. The function generator, resistor (*R*), and capacitor (*C*) model the expected behavior of the input source in the final system implementation. The variable resistance (R_L) models the load that would be attached to the output.

7.1.3 Test Case Properties

As we go through the different levels of testing we will need to build effective test cases. Effective test cases share some common attributes regardless of their level. Dianne Runnels [Run99] defined the following properties for effective test cases:

- *Accurate*. The test should check what it is supposed to and exercise an area of intent.
- *Economical*. The test should be performed in a minimal number of steps.
- *Limited in complexity*. Tests should consist of a moderate number (10–15) of steps.
- *Repeatable*. The test should be able to be performed and repeated by another person.
- *Appropriate*. The complexity of the test should be such that it can be performed by other individuals who are assigned the testing task.
- *Traceable*. The test should verify a specific requirement. The corresponding requirements for the different types of test are derived from the associated development stages in the test vee in Figure 7.1.
- *Self-cleaning*. The system should return to the pretest state after the test is complete.

7.2 Constructing Tests

This section presents the four different types of tests shown in Figure 7.1: debugging, unit testing, integration testing, and acceptance testing. This is presented in reverse order from the order in which a test should be created, as the reader is probably most familiar with basic test techniques such as debugging. Thus the presentation is from the most familiar to the more abstract. The next section presents a case study that proceeds in the opposite direction, from acceptance testing to unit testing.

7.2.1 Debugging

At some point in the design process, the implementation level is reached, where tasks such as constructing circuits, wiring integrated circuits, and writing code are carried out. Applying the functional decomposition paradigm introduced in Chapter 5 should provide a clear idea of the inputs, outputs, and behavior of the modules that are being built. Inevitably, there will come a point during the construction of a component when it will not function as expected. This is commonly referred to as a *bug*. It requires the application of debugging skills to determine the root cause of the problem and correct it. You have undoubtedly run across a variety of bugs in your day, and it is a good guess that your bugs fell into one of two camps—Bohrbugs and Heisenbugs.

Bohrbugs are named after the Bohr model of the atom that assumes that electrons have a distinct position in space. Bohrbugs are reliable bugs, in which the error is always in the same place. This is analogous to the electrons having a definite position. Given a particular input, a Bohrbug will always manifest itself in the same way and in the same place. Finding a Bohrbug is a matter of laying the correct trap. A good trap is simple to set up, quickly causes an error, and reveals the source of the error. This is a tall order, but one which experience hones.

Heisenbugs are named after the Heisenberg uncertainty principle, in which the position of an electron is uncertain. Analogously, Heisenbugs may not always be reproducible with the same input. They seemingly move around within a system and are consequently difficult to locate. Finding a Heisenbug requires you to think outside the box because they usually result from unanticipated mechanisms. An example of a Heisenbug is a computer program with a pointer error that occasionally overwrites the system stack. This can cause return values from a subroutine to be incorrect. In such a case, the subroutine would appear to have a problem, since it is returning the wrong value. However, testing the subroutine by itself would confirm that the subroutine works properly. Another good example is a circuit that works fine on some days, but doesn't work on others (typically when a professor is nearby). Insidious problems such as a floating ground line often are to blame.

Regardless of the bug type, the debugging process is iterative. You must run tests and, depending on the results, go back and run new tests. With this in mind, you should enter into the debugging process with a strategy in mind. This strategy is often similar to programming an if-then structure—"if the test is negative, then I'll pursue this line of attack; otherwise the error could be in another subsystem." In general, the debugging process is much the same as the scientific method. The steps of the debugging process are:

- *Observe.* Observe the problem under different operating conditions.
- *Hypothesize.* Form a hypothesis as to what the potential problem is.
- *Experiment.* Conduct experiments to confirm or eliminate the hypothesized source of the problem.
- *Repeat.* Repeat until the problem is eliminated.

When hypothesizing, make sure to check the simplest and easiest potential problems first. There are two good reasons for this—they are easy to perform and more tests can be performed in a given period of time. In addition, designs should be verified from the lowest levels of abstraction to the highest. For example, voltages should be verified as correct before moving to higher levels of functionality. The reason for this heuristic is obvious—the higher level of functionality cannot operate correctly unless all the lower levels are working.

7.2.2 Unit Testing

A *unit test* is a complete test of a module's functionality. In order to be a complete check, a unit test consists of a set of test cases each of which establishes that the module performs a single unit of functionality to some specification. Test cases should be written with the express intent of uncovering undiscovered defects. For example, consider a hardware module that converts an input Celsius temperature into an output Fahrenheit temperature. Let the operation of the module be represented by the following pseudocode.

```
if (16 < input < 32)
    output = ROM[input -16];
else
    output = (2 * input) + 32;
```

When the input temperature is between 16 and 32, the output is determined by a lookup operation in a ROM, otherwise the input is converted using an approximation to the familiar Celsius to Fahrenheit conversion. Each test case for this hardware module should exercise a single area of intent. Clearly, we need to have at least two test cases, one for the "if" clause and one for the "else" clause. In addition, it would be a good idea to check the boundary conditions separating the if and else clauses. Finally, we should consider the extreme values of the input. For example, if the input were a signed 8-bit number then we should check −128 and 128. If the input is a signed value then 0 is also a boundary value that should be checked.

This example illustrates the concept of a *processing path*—a sequence of consecutive instructions or states encountered from the beginning to the end of a computation process. The temperature conversion example has two processing paths, one when the if statement is taken and one when the else statement is taken. Each such processing path through the system represents a potential test case. The extent to which the test cases cover all possible processing paths is called the *test coverage*. It is desirable to design test sets that have the highest coverage possible in the fewest number of test cases. The ultimate in coverage is achieved by *path-complete coverage* where every possible path has a test. However, this level of coverage may not be possible because the number of processing paths goes up exponentially with the number of nested branches. In cases where there are more paths than it is possible to check, you must be satisfied with partial path coverage. In such cases, those paths that are thought to most likely reveal an error should be tested.

Clearly documenting unit tests has added importance because the test cases are generally written by one person or group and performed by a separate group. In order to organize the test cases they can be organized as matrices, step-by-step tests, or automated scripts.

Matrix Tests

A *matrix test* is a test that is best suited to cases where the inputs submitted are structurally the same and differ only in their values. The test procedure is then "factored out," leaving a list of inputs and their expected outputs. Since the tests are written by one group and performed by another, the test writer must leave space in the test document for the tester to make comments and observations about the system behavior.

Let's consider a test for the analog-to-digital converter (ADC) that was used in the temperature measuring system in Chapter 5 (Section 5.7, Figure 5.11). Assume that the ADC's clock frequency is 10 kHz and the input ranges from 0 to 5 V. The unit test will consist of submitting different inputs to the ADC and verifying the outputs. Since each test varies only the input, with no change in the testing procedure, the test matrix in Table 7.1 can be created.

This test case exercises each bit of the ADC's output independent of the other output bits. Other test cases should examine extreme inputs as well as illegal inputs. Care should be taken that illegal inputs do not stress the ADC beyond the manufacturer's recommendations; otherwise the tests might accidentally damage the ADC.

Table 7.1 A matrix test for an analog-to-digital converter.

Test Writer: Sue L. Engineer							
Test Case Name:	ADC unit test			**Test ID #:**	ADC-UT-01		
Description:	Verify that each bit of the output can be set independently of the other outputs.			**Type:**	☐ white box ☑ black box		
Tester Information							
Name of Tester:				**Date:**			
Hardware Ver:	1.0			**Time:**			
Setup:	Isolate the ADC from the system by removing configuration jumpers. Connect the clk input to a 10-kHz clock source and the Din input to a high-precision voltage course. Connect the output from the ADC to a logic analyzer.						
Test	V_T	Expected output		Pass	Fail	N/A	Comments
		Decimal	Hexadecimal				
1	0.000 V	0	0x000				
2	0.004887 V	1	0x001				
3	0.00977 V	2	0x002				
4	0.01955 V	4	0x004				
...	...	...	...				
10	2.502 V	512	0x200				
Overall test result:							

Step-by-Step Tests

A *step-by-step test* case is a prescription for generating the test and checking the results. These descriptions are most effective when the test consists of a complex sequence of steps. The test template for a step-by-step test has all the information contained in the matrix test template, the difference being the addition of a column in the test section describing what action the tester should perform at each step in the test process.

As an example, recall the state diagram for the vending machine in Chapter 6 (Figure 6.2) that accepts nickels and dimes and dispenses candy when a total of $0.25 (or more) is submitted. The state machine has different processing paths, depending upon the combination and order of coins deposited. Test cases can be written for each of the processing paths through the system, and an example is shown in Table 7.2 for one particular processing path.

Table 7.2 A step-by-step test for a vending machine.

Test Writer: Sue L. Engineer				
Test Case Name:	Finite State Machine Path Test #1		Test ID #:	FSM-Path-01
Description:	Simulate insertion of money with a mix of nickels and dimes. Verifies FSM outputs candy in response to a total deposit of $0.30.		Type:	☑ white box ☐ black box
Tester Information				
Name of Tester:			Date:	
Hardware Ver:	1.0		Time:	
Setup:	Make sure that the system was reset sometime prior and is in state $0.00.			

Step	Action	Expected Result	Pass	Fail	N/A	Comments
1	Strobe Nickel	State should go to $0.05				
2	Strobe Dime	State should go to $0.15				
3	Wait	State should remain $0.15				
4	Strobe Nickel	State should go to $0.20				
5	Strobe Dime	State should go to $0.25				
6	Nothing	State should go to $0.00				
	Overall test result:					

Automated Test Scripts

An *automated test script* is a sequence of commands provided to the UUT without user intervention. The outputs are usually automatically compared against the expected outputs to determine if the module contains an error. Automated scripts are executed from a device referred to by many different names such as test harness, test fixture, and test bench.

While automated scripts carry a lot of up-front cost in terms of the time required putting them together, they pay dividends when performing *regression testing*. Regression testing is the process of retesting a module after a modification in any related part of the system to ensure that no errors were inadvertently introduced. Reducing the time spent on regression testing has a positive effect on the overall development time. Hence, the benefit of automated scripts is realized later in the testing cycle. In addition, design decisions can have an effect on the amount of time spent on regression testing. It stands to reason that systems with highly coupled modules require more extensive, and consequently more time-consuming, regression testing.

The template for the matrix tests could be used to describe what an automated test script does. However, the specifics of how the automated scripts perform these actions are implementation-specific. For example, in hardware description languages the stimulus and responses of the UUT are processed by a test bench. The test bench is itself a piece of hardware coded in the same hardware language used to describe the UUT.

7.2.3 Integration Testing

After the individual subsystems have undergone their unit tests, they are then integrated into large subcomponents leading eventually to the construction of the entire system. Hence *integration testing* checks that the major modules of the overall system operate correctly together. The test cases for integration testing must be traceable to the high-level design, and the test cases are written on the basis of the characteristics of the design architecture. Test cases for integration can be derived from the following questions:

- Have all the execution paths through the system been exercised?
- Have all the modules been exercised at least once?
- Have all the interface signals been tested?
- Have all interface modes been exercised?
- Does the system meet timing requirements?

The integration tests themselves can be documented using either the matrix or step-by-step template outlined for unit tests.

7.2.4 Acceptance Testing

An *acceptance test* is a formal document stipulating the conditions under which the customer will accept the system. It generally consists of a suite of test cases that exercise the systems according to the user's environment. The test cases are constructed to ensure that the engineering requirements are met. The four attributes of a good requirement (abstract, unambiguous, traceable, and verifiable) are important in building a good acceptance test. An unambiguous requirement will result in a test that everyone can agree on. A verifiable requirement sets an objective pass/fail criterion on the acceptance test. Tests based on a traceable requirement imply they are directly assessing the needs of the project. However, an acceptance test goes far beyond an enumeration of the test cases. It typically includes the following sections:

- *Testing Approach.* The types, level, and methods employed to test the system.
- *Test Schedule.* Start and end dates for the individual tests.
- *Problem Reporting.* How the test results will be recorded.
- *Resource Requirements.* The hardware, software, and people requirements needed to perform the tests.
- *Test Environment.* The setup required to run the tests.
- *Test Equipment.* Any special equipment or configurations required to run the test.
- *Postdelivery Tests.* Tests performed on the deployed system.

- *Test Identification.* Enumeration of test cases and their unique identifiers.
- *Corrective Action.* What repairs must be made to the system in order to accept it.

It is not necessary for every test case to be passed in order for the system to be accepted. The acceptance test should stipulate the degree of importance surrounding each test. While it's easy to imagine writing the test cases for an acceptance test, the process can become a chicken-and-egg problem. That is, you are trying to stipulate the test procedures and results for a system which has yet to be implemented. This can often lead to revisions of the acceptance test plan later in the design cycle.

7.3 Case Study: Security Robot Design

In order to demonstrate the concepts involved in testing let's consider the design of a security system that monitors an office complex looking for intruders. The design team has decided to address the need by designing a mobile robot that autonomously navigates its way through the office space. The team, along with the customer, developed a number of requirements, and from this we will focus on two that address a fundamental navigational problem.

- *The robot's center must stay within 12 to 18 centimeters of the wall over 90% of the course, while traveling parallel to a wall over a 3 meter course.*
- *The robot's heading should never deviate more than 10 degrees from the wall's axis while the robot travels parallel to a straight wall over a 3 meter course.*

This case study explores test cases for the acceptance, integration, and unit testing related to these two requirements. The development of the test cases follows the proper order of test case development illustrated in Figure 7.1. That means acceptance tests are developing in conjunction with the requirements, integration tests are constructed during the system design, and unit tests are constructed during the system build.

Acceptance Testing

We start by constructing an acceptance test case to verify that the robot can achieve the stated requirements. A number of tests would need to be built, and we create a test only for the first engineering requirement. A test could be performed by having someone observe the robot moving along a wall and mark (on the floor) whenever the robot strayed out of bounds. Such a test would not easily be repeatable because different people might judge what is meant by "out of bounds" differently. The accuracy of such a test is questionable because determining when and if a speeding robot crossed the boundary is difficult. Finally, there would be no permanent record of the test results making it difficult for the customer to actually verify the test was passed. A way to address these problems is to have the robot monitor its own distance from the wall. This is done in this case by a program written to monitor the position of the robot over time and store these values. From the specifications for a step-by-step acceptance in Table 7.3 it is clear that the test program must configure the robot to log the distance data while traversing the wall. After this data is downloaded from the robot, it can be analyzed in a spreadsheet program to determine the needed metrics and archived for future reference.

Table 7.3 A step-by-step acceptance test case for the autonomous robot.

Test Writer: Sue L. Engineer							
	Test Case Name:	Robot acceptance test #1			Test ID #:	Robot-AT-01	
	Description:	Checks the engineering requirement: *The robot's center must stay within 12 to 18 centimeters of the wall over 90% of the course, while traveling parallel to a wall over a 3-meter course.*			Type:	☐ white box ☑ black box	
Tester Information							
	Name of Tester:				Date:		
	Hardware Ver:	Robot 1.0			Time:		
	Setup:	Completed robot should be fully charged and placed on 3-meter test track.					
Step	Action	Expected Result	Pass	Fail	N/A	Comments	
1	Write a program to monitor the robots' position from the wall.	Program should be statically tested to verify accuracy. Should sample wall at a sufficient rate, depending on speed.					
2	Put robot on test track, run test, and download data.	The robot should travel down the entire length of the test track and then stop.					
3	Plot test data in a spreadsheet program.	Plot of position vs. time should be within 12–18 cm 90% of the time.					
	Overall test result:						

Integration Testing

The team, in consultation with the customer, has gone through the requirements and created a complete set of acceptance tests in addition to the test in Table 7.3. They next turn to developing a high-level design architecture that can meet the requirements. The design they create is shown in Figure 7.4, the Level 1 architecture of the autonomous robot.

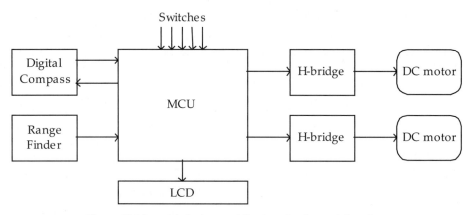

Figure 7.4 Level 1 design architecture for the mobile robot.

The heart of the design is a microcontroller (MCU) that reads the sensor values, makes decisions, and controls the speed of the two drive motors. The robot moves and turns by adjusting the relative speed of each motor, using a pulse-width-modulated (PWM) signal from the MCU. The duty cycle of the PWM is directly proportional to the speed of the motor. The H-bridges then amplify the MCU output to power to the motors. The MCU also outputs a set of signals to send text to an LCD. The signal to the digital compass is bidirectional because the MCU must configure the operating mode of the compass before using it. The MCU receives an analog input from the range finder, where the voltage level of the signal is proportional to the distance to the obstacle. Finally, a set of switches is included that allows for manual input and testing of the robot.

Clearly, the interaction of the MCU with each of the compass, rangefinder, LCD, switches, and H-bridges should be examined. However, this will be left to the unit test because the MCU makes a great test harness that can be used to provide stimulus to and read outputs from these I/O. In addition, many of the routines from these tests can be reused later in the development process.

There are many interactions between subsystems that could be tested during integration testing. A careful examination of the system must be done to determine which combinations of subsystems are most likely to create problems. Experience and component selection play a large part in molding expectations. For example, the magnetic field created by the windings in the DC motor can affect the reading generated by the compass. This interaction could potentially affect the headings read by the MCU and cause the robot to go off course by more than the allowable 10 degrees. Thus a step-by-step integration test is created in Table 7.4 to test the operation of the motors with a magnetic compass.

Table 7.4 A step-by-step integration test case for the compass and motors.

Test Writer: Sue L. Engineer				
Test Case Name:	Robot integration test #1	**Test ID #:**	Robot-IT-01	
Description:	Checks interaction of DC motors on the magnetic compass.	**Type:**	☐ white box ☑ black box	
Tester Information				
Name of Tester:			Date:	
Hardware Ver:	Robot 1.0		Time:	
Setup:	A wooden turntable should be placed on top of the cardinal direction map. This map should be aligned with a magnetic compass. There should be no metal present while the alignment is being performed. Next, the partially assembled robot should be placed on the turntable. The MCU should be connected to a terminal to observe and record data.			

Step	Action	Expected Result	Pass	Fail	N/A	Comments
1	Write program to spool compass readings while simultaneously driving motors.	Program should be statically tested to verify accuracy. Should sample compass at a sufficient rate, depending on speed.				
2	Run acceptance test.	Test program should prompt user to turn the robot to an orientation and then run the motors up to full speed.				
3	Plot spooled data in spreadsheet program.	Plots should be analyzed to see if compass deviated any more than 10 degrees from set point.				
	Overall test result:					

It is clear from this test case that a testing program needs to be written in order to prompt the user to align the robot, capture compass readings, and then to spool them back to the user. The requirement that the compass readings deviate no more than 10 degrees is based on the engineering requirement that the robot deviate no more than 10 degrees from the wall's axis while navigating down the hallway.

Unit Testing

Once the Level 1 architecture is developed and the test cases written to ensure that the architecture is capable of meeting the design requirements, the design team moves on to selecting components to use in the design. The design team must select the units so that the resulting system can meet the engineering requirements. Each of the individual components in Figure 7.4 needs to be considered as a candidate for unit testing. In general each functional unit might have several test cases constituting its unit test. A unit test for the digital compass component will illustrate this. Before the test case is presented, note that the functional design requirements for the unit are as given in Table 7.5.

Table 7.5 The functional requirements for the digital compass.

Module	Digital Compass—Geosensor version 2.3
Inputs	- Earth's magnetic field: An orientated field of magnetic force beginning and ending at the earth's magnetic poles. - SClk—Clock signal to clock data through the module. Maximum frequency is 10 Mhz. - SDIn—Serial data input to send data into the compass module. Date is valid on positive SClk edges.
Outputs	- SDOut—Serial data output from the compass module. Data is valid on negative clock edges.
Functionality	Senses the earth's magnetic field and determines the orientation of the compass with respect to the field. This orientation is stored in an internal register and can be retrieved through the SPI interface.
Test	Comp-UT-01

In order to be useful in the overall design the compass module must be able to accurately report the robot's heading. The requirements place an upper bound of 10 degrees on the error in the heading of the robot. Thus, the matrix test case in Table 7.6 is constructed to configure the compass and then reads heading data from it. This unit test looks for heading errors greater than 10 degrees.

Table 7.6 Matrix unit test for the digital compass.

Test Writer: Sue L. Engineer			
Test Case Name:	Compass unit test #1	**Test ID #:**	Comp-UT-01
Description:	Checks that the compass returns correct angular measurements to the MCU. Test program is in ./test/compass_unit_test_1.c	**Type:**	☐ white box ☑ black box
Tester Information			
Name of Tester:		**Date:**	
Hardware Ver:	Compass Module - Geosensor version 2.3	**Time:**	
Setup:	Compass module should be wired to the MCU through the SPI interface pins. The MCU should be connected to an RS232 terminal through its SCI interface. The terminal should be configured to run at 9600 baud. Cardinal directions map should be aligned using the magnetic compass.		

Step	Action	Expected Result	Pass	Fail	N/A	Comments
1	Compile compass.c in /test directory	IDE should generate no warnings or errors.				
2	Download	MCU should report "download successful."				
3	Execute	MCU should display compass splash screen on terminal interface.				
4	Orientate compass to 0 degrees.	Terminal interface should display 0 degrees +/− 10 degrees.				
5	Orientate compass to 30 degrees.	Terminal interface should display 30 degrees +/− 10 degrees.				
6	Orientate compass to 45 degrees.	Terminal interface should display 45 degrees +/− 10 degrees.				
…	…	…				
12	Orientate compass to 315 degrees.	Terminal interface should display 315 degrees +/− 10 degrees.				
	Overall test result:					

7.4 Guidance

Tests have a lifetime beyond the obvious need to check proper operation of the subsystems, their integration, and the overall performance of the system. Test cases describe how the system operates in plain English. Test cases can be used to develop diagnostics, assist in writing technical documentation, and aid marketing and sales staff in understanding system performance. Testing is a value-added process in design. Beyond attempting to remove bugs from the system, Burke and Coyner [Bur03] suggest the following are good reasons to perform testing:

- *Testing reduces the number of bugs in existing and new features*. Testing does not eliminate all the bugs, but rather reduces the probability of a bug making it to production.
- *Tests are good documentation*. Tests provide insight to others on the operation of the unit under test and how to interface to it.
- *Tests reduce the costs of change*. A change to a complex design with no tests can produce bugs that are difficult to track down. A good set of regression tests can help localize the effect of bugs introduced by changes.
- *Tests improve design*. In order to create a testable design, you need to create highly cohesive, loosely coupled units.

- *Tests allow you to refactor*. Subcomponents of a testable design can be changed and optimized with less chance of introducing new errors. This is because tests exist that can verify the redesigned (refactored) module functions correctly.
- *Tests constrain features*. When a test is written before building the associated module, the exact requirements are defined. Hence, when a unit passes its test, there is confidence that the requirements have been met.
- *Tests defend against other designers*. Often a design needs to have circuitry to deal with special cases. Tests that check these special cases can make sure that future modification do not remove them.
- *Testing is fun*. Writing tests requires creative solutions to complex design problems.
- *Testing forces you to slow down and think*. When writing a test before incorporating a feature into a design, you are forced to see how the new feature fits into the existing design framework.
- *Testing makes development faster*. On a component level, testing slows development. However, as the design becomes larger and more complex, modules can be more easily integrated into the design without causing malfunctions in existing components.
- *Tests reduce fear*. Would you rather improve a unit with a test suite or one without?

7.5 Summary and Further Reading

Testing is an important part of the design process that helps to ensure systems will operate properly. This chapter examined basic principles of testing including black box testing, white box testing, controllability, and observability. They address the manner in which tests can be conducted, controlled, and states of the system observed. The use of stubs, which are employed to simulate system inputs and outputs were examined, as well as the properties of test cases. The different phases of testing from unit tests through integration tests to acceptance tests were examined. Testing proceeds from the most detailed level of the system to the most general, and the tests performed in each phase are traceable to their corresponding phases in the design development process.

The field of testing has been well developed by the software engineering community. Software Engineering: An Engineering Approach [Pet00] provides a good overview of testing principles such as black box and white box testing. It also includes a number of test strategies beyond those considered here. The Glossary of Vulnerability Testing Terminology from the University of Oulu's Electrical and Information Engineering Department [Oul04] provides an extensive list of terms related to testing in the software domain. Gray's 1985 article "Why Do Computers Stop and What Can Be Done About It?" [Gra85] coined the terms Heisenbug and Bohrbug. This article introduces many interesting facts about how supercomputers fail. It provides a rare chance to look at the inner world of a supercomputer company. Many of the topics in the unit test section were influenced by Dianne L. Runnel's article "How to Write Better Test Cases" [Run99]. In this article, she defines precisely what is meant by unit test, and gives a clear picture of how to construct a unit test. This article, along with many other scholarly articles on testing, can be found at *www.stickyminds.com*. An exceptional set of documents and templates are available from the Systems Engineering Processing Group of the United States Air Force. While intended for software development, the checklists for unit and integration testing contain many insightful points. They are accessible at *https://ossg.gunter.af.mil/applications/sep/menus/Main.aspx*. The list of acceptance test items was due in part to the information found at *http://www.tbs-sct.gc.ca/emf-cag/acceptance/outline/atpo-vper_e.asp*.

7.6 Problems

7.1 Explain the differences between black box and white box testing.

7.2 Identify a circuit simulator (analog or digital) that you are familiar with. Explain the features of this simulator that increase the observability and controllability of the circuit being simulated.

7.3 A mobile robot is being built. It uses two DC motors in a differential drive configuration: a microcontroller to control movement and an ultrasonic sensor to detect obstacles. The robot is built to wander around without bumping into objects. Explain how stubs could be used in testing to take the place of incomplete subsystems. Be specific.

7.4 Consider that you have an op amp integrated circuit package, such as the LM741 in Appendix C. What type of testing would be appropriate for testing this device? Write a short test plan for doing so.

7.5 Explain under what situations a matrix test is appropriate.

7.6 Explain under what situations a step-by-step is test appropriate.

7.7 Consider the stages of unit testing, integration testing, and acceptance testing. For each of these stages, identify the corresponding requirements that each test should be traceable to.

7.8 Consider the case study robot design in Section 7.3, which presents an acceptance test for the first system requirement. Develop an acceptance test for the second system requirement.

7.9 Consider the case study robot design in Section 7.3. Develop an integration test that demonstrates the combined operation of the DC motors, MCU, and range finder.

7.10 Consider the case study robot design in Section 7.3. Develop an integration test that demonstrates the combined operation of the digital compass, MCU, and LCD.

7.11 Consider the case study robot design in Section 7.3. Develop unit tests for the rangefinder, the DC motors, the H-bridges, and the LCD.

7.12 **Project Application.** Develop an acceptance test suite for your project. The acceptance tests should apply to the engineering requirements developed for the system.

7.13 **Project Application.** Develop an integration test suite for your project. The integration tests should apply to the higher levels of the design architecture and address the interaction between functional units.

7.14 **Project Application.** Develop a unit test quite for your project. The unit tests should apply to the lowest level units in the design.

Chapter 8 System Reliability

Quality is never an accident. It is always the result of intelligent effort. —John Ruskin

A typical design project in your academic career may never leave the confines of a laboratory. However, in industry, engineers develop systems that are used by the public at large, and issues beyond the functionality, such as reliability, safety, and maintainability, become important factors in the success of the design. Over the past 20 years, industry has made a great shift to address reliability through the adoption of processes such as quality functional deployment (QFD), six sigma (Figure 8.1), and robust design. While other chapters have addressed some elements of these processes, the objective of this chapter is to examine system reliability. Reliability attempts to answer the question of how long a system will operate without failing. Answering this question has inherent uncertainty and requires the use of probability and statistics. This chapter presents a review of basic probability theory and applies it to estimate the behavior of real-world devices. Reliability at the component and system levels is considered.

Learning Objectives

By the end of this chapter, the reader should:
- Have a familiarity with the basic principles of probability and understand how they apply to reliability theory.
- Understand the mathematical definition and meaning of failure rate, reliability, and mean time to failure.
- Understand how to determine the reliability of a component.
- Understand how to derate the power of electronic components for use under different operating temperatures.
- Understand how to determine the reliability of different system configurations.

DILBERT® by Scott Adams

Figure 8.1 Dogbert's six sigma program. (Dilbert © Scott Adams / Dist. by United Feature Syndicate, Inc.)

8.1 Probability Theory Review

Probability theory provides a formal framework to study chance events. It is a powerful tool for modeling engineering systems and is a requisite for reliability estimation. Although this section provides a review of some important concepts from probability, it is assumed that the reader is versed in the basics of probability theory.

In order to apply probability, some general definitions are examined first. An *experiment* is the process of measuring or quantifying the state of the world. The particular outcome of an experiment is an *event* (e_i), while the *event space* (E) is the set of all possible outcomes of the experiment. For example, consider an experiment where a six-sided die is rolled. The experiment is rolling the die and observing the outcome, the event is the particular outcome observed, and the event space for the experiment is the set $E = \{1, 2, 3, 4, 5, 6\}$. The outcomes do not have to be numerical values. Another example experiment is tossing a coin, in which case the event space is $E = \{heads, tails\}$. Both are examples of a discrete event space because there is a finite number of experimental outcomes. In a discrete event space, the union of all the possible experimental outcomes defines the event space. If e_i is the ith event in a discrete event space, then the event space is given by the union

$$E = \bigcup e_i. \tag{1}$$

The probability of an event indicates how likely it is for an event to occur. This is quantified by the probability operator $P()$, which assigns to each event a real number between 0 and 1. The probability is the percentage of times that an event would occur if the experiment were repeated an infinite number of times (the law of large numbers). Two of the three fundamental axioms on which probability theory is built are

$$P(e_i) \geq 0 \tag{2}$$

$$P(E) = 1. \tag{3}$$

The first axiom indicates that all probabilities are nonnegative, while the second is a restatement of the event space definition—the outcome of an experiment must be an element of the event space. Armed with these definitions and axioms, we now examine some important concepts from probability.

8.1.1 Probability Density Functions

Not all event spaces are discrete as in the case of rolling a die or flipping a coin. Consider an experiment where the objective is to measure temperature. Clearly, such a measurement requires a variable having a continuous range of possible values. A random variable is defined as the outcome of an experiment that has a continuum of possible values. Random variables have a mathematical function known as the *probability density function* (PDF) associated with them, which when integrated, yields the probability of a range of events. A PDF is typically denoted as $p_X(x)$, where X takes values over the event space. Standard notation identifies random variables by using uppercase variables as the subscript for the PDF. The variable inside the parentheses is a lowercase dummy variable that does not have to match the random variable, but typically does. A question that the PDF allows us to ask is "What is the probability that a random variable is in some range?" Consider the case where the objective is to determine the probability that the random variable X lies between two values a and b. Written with the probability operator, this is indicated as $P(a \leq X \leq b)$. It is determined from the PDF as follows:

$$P(a \leq X \leq b) = \int_a^b p_X(x)\, dx. \qquad (4)$$

Conceptually, this probability represents the area under the PDF between the two limits of integration as shown in Figure 8.2.

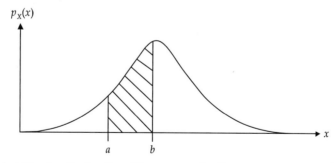

Figure 8.2 A probability density function. The area under the curve represents the probability that the random variable X lies in the interval [a, b].

Let's examine a few more important properties of probability density functions. The first, which is analogous to (3), indicates that the probability of the event space occurring is equal to one. This is known as the normalization property, and it is expressed as

$$\int_{-\infty}^{\infty} p_X(x)\, dx = 1. \tag{5}$$

Another interesting result is obtained by trying to determine the probability that a random variable takes on an exact value, for example $P(X = a)$. That is determined from the integral

$$P(X = a) = \int_a^a p_X(x)\, dx = 0. \tag{6}$$

This is a somewhat counterintuitive result—it indicates that the probability a random variable can take on a particular value is zero. Does this make any sense? Consider an experiment where the objective is to measure a voltage value for a random variable V. Now consider the question, "What is the probability that the result of a voltage measurement equals π (the irrational number) volts?" In practice, this question is impossible to answer because the precision required of the meter is infinite and contrary to its construction, so the mathematical and practical results are in harmony. There is a way around this dilemma, which is to determine the probability that the random variable is within a small range about the target value, as follows:

$$P(\pi < V < \pi + \Delta v) = \int_{\pi}^{\pi + \Delta v} p_V(v)\, dv \approx p_V(\pi)\, \Delta v. \tag{7}$$

This means that the probability a random variable is within a small range about a given value is approximated by the product of the PDF evaluated at the value and the size of the range.

8.1.2 Mean and Variance

Two useful and well-known statistics that are determined from the PDF are the mean μ and variance σ^2. They are found from the PDF as follows:

$$\mu_X = \int_{-\infty}^{\infty} x p_X(x)\, dx \tag{8}$$

$$\sigma_X^2 = \int_{-\infty}^{\infty} (x - \mu)^2 p_X(x)\, dx. \tag{9}$$

The *mean* is analogous too the center of mass of the PDF; it is also known as the average value. The *variance* is the average of the squared difference between the mean and the values of the PDF, where the squared term ensures that a positive difference is taken. The square root of the variance is known as the standard deviation σ.

8.1.3 Common Probability Density Functions

There are many PDFs available for describing the seemingly random variations in the behavior of observed systems and phenomena. In this section, three common PDFs (normal, exponential, and uniform) are presented.

The Normal Density

The most common density function encountered in the physical sciences and engineering is the normal density. Many population variations can be described by a normal density. For example, the resistance values of a large batch of 2.2 kΩ resistors would likely follow a normal density. The normal density is defined as

$$p_X(x) = \frac{1}{\sqrt{2\pi}\sigma} e^{-\frac{1}{2}\left(\frac{x-\mu}{\sigma}\right)^2}. \tag{10}$$

The mean μ and standard deviation σ are part of the definition of the PDF and used to alter the shape of the density to suit the particular need. The normal PDF is plotted in Figure 8.3. Varying μ allows the overall function to be shifted along the x axis, while increasing σ spreads (or flattens) the function out. Calculating probabilities from the normal density can be done (although it takes a bit of work mathematically) so they are usually computed from something known as the cumulative distribution function, which is presented shortly.

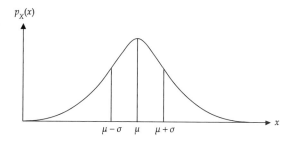

Figure 8.3 A normal density function with the mean μ and standard deviation σ shown.

The Uniform Density

The uniform density, plotted in Figure 8.4, models the outcome of an experiment where all outcomes are equally likely. Mathematically, the PDF for a uniform density is given by

$$p_X(x) = \frac{1}{b-a}, \quad a \leq x \leq b, \tag{11}$$

where a and b are selected to meet the demands of a particular problem.

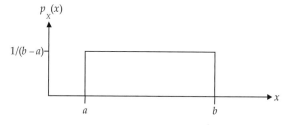

Figure 8.4 The uniform density on the interval [a, b].

The Exponential Density

Exponential densities are often utilized to model time-dependent functions, such as interarrival times between data packets in communication systems. As shown later, the exponential density also describes the behavior of component failures as a function of time. The mathematical description of an exponential density is

$$p_X(x) = \lambda e^{-\lambda x}, \quad x \geq 0, \lambda \geq 0. \tag{12}$$

The PDF is characterized by the parameter λ, which affects the shape of the curve as demonstrated in Figure 8.5.

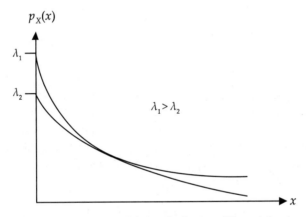

Figure 8.5 The exponential density for two different λ values.

8.1.4 Cumulative Distribution Functions

An important class of questions can be phrased as, "What is the probability that a random variable X is less than value a?" For example, the objective might be to determine the probability that an electronic component will malfunction within 2 years. Returning to the first question, it is clear that the goal is to determine the probability $P(X < a)$, which is found by integrating the PDF. This result is generalized by allowing the upper limit of integration to take on an arbitrary value that spans the range of the random variable. This produces a new function, known as the *cumulative distribution function* (CDF), which is the integral function of the PDF and is defined as

$$\text{CDF}(x) \equiv \int_{-\infty}^{x} p_X(y) \, dy. \tag{13}$$

8.2 Reliability Prediction

Our main interest in the study of probability stems from the desire to quantify the reliability of a system. The following is a formal mathematical definition of *reliability*.

Definition: Reliability, $R(t)$, is the probability that a device is functioning properly (has not failed) at time t.

In order to determine $R(t)$, it is necessary to first introduce some related mathematical entities and their meanings. The *failure rate*, $\lambda(t)$, of a device is the expected number of failures per unit time. The failure rate is measured by operating a batch of devices for a given time interval and noting how many fail during that interval. A typical graph of failure rate versus time has the bathtub shape shown in Figure 8.6. The high initial failure rate is a result of manufacturing defects and is often referred to as infant mortality. Consequently, many manufacturers will "burn-in" devices at the factory, so that if they fail, they do so before being sold. After the infant mortality phase, devices enter a phase of constant failure rate, where $\lambda(t) = \lambda$, known as the service life. Estimates for λ are determined empirically by testing a large number of components. They are usually expressed as a unit failure per a given number of hours, for example $\lambda = 1$ failure/10^6 hours. After some period of time, devices start to wear out and the failure rate increases. This usually happens as a result of mechanical wearing with age and use. Properly designed electronic devices will not have a wear-out region, instead continuing on at a constant failure rate. This applies only to the electronic devices themselves, not necessarily to complete systems that will likely contain mechanical devices.

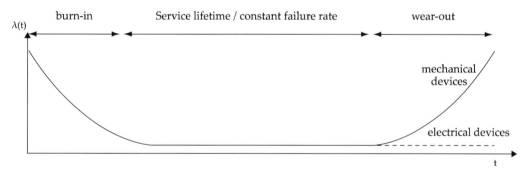

Figure 8.6 Failure rate as a function of time, also known as the bathtub curve.

In addition to failure rate, a PDF for the *failure time* of the device, $f_T(t)$, is defined, where the random variable is time T. This function allows the question to be asked "What is the probability that a device will fail between time t_1 and t_2?" It is important to note the difference between $\lambda(t)$ and $f_T(t)$. The failure rate tells us the average rate that a collection of identical devices will fail at a given time t, while $f_T(t)$ is a PDF used to determine the probability that a given device will fail within a specified time period. A CDF for $f_T(t)$, is determined as

$$F(t) = \int_0^t f_T(\tau)\, d\tau. \tag{14}$$

$F(t)$ answers the question "What is the probability that the device has failed by time t?" and it is also known as the *failure function*. Take a few seconds to go back and review the definition of $R(t)$. It is clear that $R(t)$ is directly related to $F(t)$ and is its complement. The relationship between the two is

$$R(t) = 1 - F(t). \tag{15}$$

Since $F(t)$ is a CDF, it increases monotonically from an initial value of 0 to a maximum value of 1 as time goes to ∞ as shown in Figure 8.7. Conversely, $R(t)$ starts at a value of 1 at time zero and decreases monotonically to a value of 0.

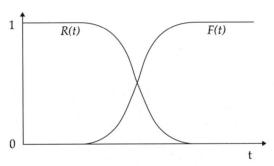

Figure 8.7 Example reliability and failure functions.

Since $\lambda(t)$ represents data that is measured empirically, it is useful to establish a relationship between $\lambda(t)$ and the ultimate goal of reliability, $R(t)$. To do so, a relationship between $\lambda(t)$, $R(t)$, and $f_T(t)$ is established as follows. Consider a small period of time between t and $t + \Delta t$, and determine the probability of device failure during this period. From the approximation developed in (7), this probability is given by

$$P(\text{failure between } t \text{ and } \Delta t) \approx f_T(t)\, \Delta t. \tag{16}$$

How is this probability related to $R(t)$ and $\lambda(t)$? $R(t)$ provides the probability that the device is working at time t and $\lambda(t)$ gives the probability that the device will fail at time t. The product of $R(t)$, $\lambda(t)$, and Δt gives the same probability of failure in (16):

$$P(\text{failure between } t \text{ and } \Delta t) = R(t)\lambda(t)\, \Delta t. \tag{17}$$

Equating (16) and (17) provides the desired relationship between the three quantities,

$$f_T(t) = R(t)\lambda(t), \tag{18}$$

that is fundamental in establishing the connection between $R(t)$ and $\lambda(t)$. However, the PDF $f_T(t)$ needs to be eliminated from (18). This is accomplished through its relationship to the CDF $F(t)$, and thus $R(t)$, as follows

$$f_T(t) = \frac{d}{dt}F(t) = \frac{d}{dt}[1-R(t)] = -\frac{d}{dt}R(t). \tag{19}$$

Equating this result with (18) produces

$$-\frac{d}{dt}R(t) = R(t)\lambda(t) \quad \Rightarrow \quad \frac{-\frac{d}{dt}R(t)}{R(t)} = \lambda(t). \tag{20}$$

Integrating both sides gives

$$\int_0^t \left(\frac{-\frac{d}{d\tau}R(\tau)}{R(\tau)}\right)d\tau = \int_0^t \lambda(\tau)\,d\tau \quad \Rightarrow \quad -\ln(R(t)) = \int_0^t \lambda(\tau)\,d\tau \tag{21}$$

and solving for $R(t)$ produces the final result for reliability as a function of $\lambda(t)$,

$$\boxed{R(t) = \exp\left[-\int_0^t \lambda(\tau)\,d\tau\right]}. \tag{22}$$

During the service lifetime phase, the failure rate is constant, simplifying equation (22) to

$$\boxed{R(t) = \exp(-\lambda t)}. \tag{23}$$

This important result is now applied in Example 8.1.

Example 8.1 Transistor Reliability.

Problem: Consider a transistor with a constant failure rate of $\lambda = 1/10^6$ hours. What is the probability that the transistor will be operable in 5 years?

Solution: This solution is found using the reliability function for a constant failure rate in (23) as follows.

$$R(t) = \exp(-\lambda t)$$

$$R(5 \text{ years}) = \exp\left(-\frac{1}{10^6 \text{ hours}} \times \frac{24 \text{ hours}}{\text{day}} \times \frac{365 \text{ days}}{\text{year}} \times 5 \text{ years}\right)$$

$$= \exp(-0.0438)$$
$$= 0.957$$
$$= \underline{95.7\%}.$$

8.2.1 Mean Time to Failure

The *mean time to failure* (MTTF) is a quantity that answers the question, "On average how long does it take for a device to fail?" From its definition, it is apparent that the MTTF is the mean value of the random variable T (failure time). It is determined from the PDF and the definition of the mean in (8) as follows

$$\text{MTTF} = \int_0^\infty t f_T(t)\, dt. \tag{24}$$

At this point the form of the PDF for $f_T(t)$ is not known, but it can be found from (19) since it is the negative derivative of $R(t)$. Assuming the form of $R(t)$ found in (23) for a constant failure rate gives

$$f_T(t) = -\frac{d}{dt}R(t) = \lambda e^{-\lambda t}. \tag{25}$$

This means that under the condition of a constant failure rate, the failure PDF follows an exponential density. The MTTF is found from $f_T(t)$ via integration by parts to be

$$\boxed{\text{MTTF} = \int_0^\infty t e^{-\lambda t}\, dt = \frac{1}{\lambda}.} \tag{26}$$

This makes intuitive sense because λ is the expected number of failures per unit time for a device. Consequently, the reciprocal of λ is the expected time between failures or MTTF. Let's consider a few examples.

Example 8.2 Transistor MTTF.
Problem: Consider the transistor in Example 8.1. Determine (a) the MTTF, and (b) the reliability at the MTTF.
Solution:
(a) From (26),

$$\text{MTTF} = \frac{1}{\lambda} = \frac{1}{(1/10^6 \text{ hours})} = 10^6 \text{ hours}$$

$$= \underline{114 \text{ years}}.$$

(b) From (23) the reliability at 114 years is

$$R(t) = \exp(-\lambda t)$$

$$R(114 \text{ years}) = \exp\left(-\frac{10^6 \text{ hours}}{10^6 \text{ hours}}\right) = \exp(-1) = 0.368$$

$$= \underline{36.8\%}.$$

This is a bit counterintuitive. Although the average time between transistor failures is 114 years, an individual transistor has only a 36.8% chance of surviving to 114 years. It would seem logical that the reliability at 114 years should be 50% and that the transistor would have a 50-50 chance of failing. This would be true if $f_T(t)$ were symmetric about its mean, but that is not the case for the exponential density.

Example 8.3 Human lifespan estimation.

Problem: Data shows that for a 30-year-old population, the failure (death) rate is constant with approximately 1.1 deaths per 1000 people per year. Given this data, estimate the MTTF of humans.

Solution: In order to find MTTF, λ is needed. From the information given it is

$$\lambda = \frac{(1.1/1000) \text{ failures}}{1 \text{ year}} = \frac{1.1 \text{ failure}}{10^3 \text{ years}} = \frac{1 \text{ failure}}{909 \text{ years}}$$

From this MTTF is computed as

$$\text{MTTF} = \frac{1}{\lambda} = 909 \text{ years!}$$

While great news for those of us seeking longevity, this calculation is clearly wrong since the upper limit on human lifespan is empirically known to be about 120 years. Why is this so? Serious problems arise if $R(t)$ is used in situations where the underlying assumption is invalid. The results in (23) and (26) apply only if the failure rate is constant. Although that is nearly true for people in their 20s and 30s, it is not true as people age. People do wear out and the failure rate increases with age.

8.2.2 Failure Rate Estimates

The overriding objective of this chapter is to estimate the future behavior of devices that are used in electrical and computer systems. The particular behavior of interest is the state of a device's functionality—the reliability, is it working or has it failed? Equation (23) indicates that it is fairly straightforward to determine reliability, if the failure rate (λ) is known and is constant. One question to consider is what factors influence the failure rate of a device. Many of us probably have had experiences in the laboratory where we have caused devices to fail by subjecting them to conditions outside of the normal operating bounds, notably excessive current, power, or heat. In those cases the devices probably failed, or burned up, as a result of operating conditions outside the allowed bounds for the device. However, even when operated within the allowable norms of a device's operating conditions, variations in factors such as power, operating voltages, and temperatures affect λ.

The United States Military has kept copious records of device failures in the field and the conditions under which the devices operated. These records are synthesized in a handbook entitled Reliability Prediction of Electronic Equipment [MIL-HDBK-217F] that provides failure rates for various analog and digital components, along with adjustment factors to account for operating conditions, the environment, and device quality. Categories of devices included in the handbook are switches, fuses, diodes, optoelectronic devices, and microelectronic devices (op amps, logic devices, microcontrollers, microprocessors). The handbook was last published in 1991 and has been discontinued, but it is still widely accepted and used. Bellcore (subsequently Telcordia) has developed newer models [Tel96] based upon MIL-HDBK-217F that were updated to better predict the reliability of components. MIL-HDBK-217F is used here since it is freely available in the public domain.

Failure rates for resistors, capacitors, transistors, and integrated circuits from MIL-HDBK-217F are included in Appendix C. For each device, a base failure rate is given, λ_b, and multiplied by a number of adjustment factors, denoted by the symbol π, to estimate the device failure rate λ. Each adjustment factor has a unique subscript, and the factors' values are found from tables or equations in the handbook.

For example, consider the low-frequency field-effect transistor in Appendix C. The overall failure rate is given by the equation $\lambda = \lambda_b \pi_T \pi_A \pi_Q \pi_E$ failures/10^6 hours. λ_b is the base failure rate that is directly read from a table, π_T is a temperature factor that is computed from an exponential equation (be careful to use the junction temperature as indicated in Appendix C), π_A is an application factor that depends upon how the device will be used, is a quality factor, and π_E is an environmental factor. The quality factor table lists some strange names and values from 0.7 to 8.0. The quality factor describes the level of burn-in and screening each device receives before leaving the factory. Joint Army/Navy (JAN, JANTX, and JANTXV) quality factors are the highest standard, and are usually required only for space vehicles. That individual attention to burn-in means that JAN parts are expensive and most JAN devices have passed their infant mortality phase before leaving the factory. In determining failure rate for a device from a table, it is common practice to always round parameters or values pessimistically so that the evaluation is a worst-case analysis of its performance. That way the device should perform with a higher reliability when embedded into a system, hopefully causing only pleasant surprises in operation. Finally, the factor π_E is based upon the different operating environments that are identified in Appendix C. Example 8.4 demonstrates the application of the MIL-HDBK-217F standard for reliability estimation.

In summary, it is possible to estimate the reliability of devices if the failure rate is known and it is constant. The U.S. Military and Telcordia handbooks provide guidance for estimating failure rates, and thus the component reliability. It must be kept in mind that they are estimates and not guaranteed to predict the exact performance. It is also apparent that this can become a rather time-consuming process if there are many components in a system, and thus the use of reliability software packages may be warranted.

Example 8.4 Reliability Estimation Using the MIL-HDBK 217F.

Problem: Consider the circuit below that contains a bipolar junction transistor (BJT). This electronic circuit is a simple digital logic inverter. When the input voltage V_1 is 0, the BJT is off, no current flows through any branches of the device, and the output voltage, $V_{0,}$ is 5 V. When the input is 5 V (high) the BJT goes into saturation because of the high base current (low R_B), producing a 50 mA collector current, a large voltage drop across R_C, and an output voltage close to 0 V. The average collector current for the two states is 25 mA, producing an average power of 125 mW (25 mA × 5 V). Assume that the circuit is used in a missile launcher, the ambient temperature is 25°C, and that JANTX-quality parts are used. Determine the MTTF and reliability for the 2N3904 BJT (a low-power, low-frequency BJT) in 20 years.

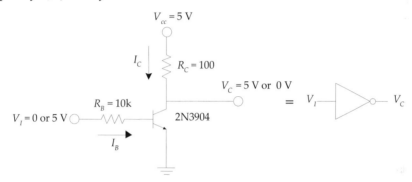

Solution: The objective of this problem is to determine the failure rate, from which the MTTF and reliability are estimated. From the MIL-HDBK-217F data in Appendix C, the failure rate is

$$\lambda = \lambda_b \pi_T \pi_A \pi_R \pi_S \pi_Q \pi_E \; \frac{\text{failures}}{10^6 \text{ hours}}$$

The base failure rate is given directly as $\lambda_b = 0.00074$. π_T is the temperature factor and its value is determined from the relationship

$$\pi_T = \exp\left[-2114\left(\frac{1}{T_J + 273} - \frac{1}{298}\right)\right].$$

where T_J is the junction temperature. As indicated in Appendix C, it is computed as

$$T_J = T_A + \theta_{JA} P_D$$

$$= 25°C + \left(200 \frac{°C}{W}\right)(125 \times 10^{-3} \text{ W}) = 50°C.$$

The thermal resistance, θ_{JA}, is read from the 2N3904 datasheet (Appendix D) and will be examined in more detail shortly. The temperature factor is

$$\pi_T = \exp\left[-2114\left(\frac{1}{50+273} - \frac{1}{298}\right)\right] = 1.73.$$

π_A is an application factor (switched or linear amplification), and the value for the switched logic inverter is $\pi_A = 0.70$. π_R is a power rating factor that is computed based upon the maximum rated power dissipation of the 2N3904 (625 mW from the component datasheet in Appendix D) as follows:

$$\pi_R = (P_R)^{0.37} = (0.625)^{0.37} = 0.84.$$

π_S is a stress factor that is computed from the ratio of the maximum collector-emitter voltage over the maximum rated value of the device. In this circuit, the maximum value of V_{CE} is 5 V (when the device is off and no current flows), while the maximum rated value of V_{CE} from the 2N3904 datasheet is 40 V.

$$V_S = \frac{\text{applied } V_{CE}}{\text{rated } V_{CE}} = \frac{5}{40} = 0.125$$

$$\pi_S = 0.045 \exp(3.1 \times V_S) = 0.045 \exp(3.1 \times 0.125) = 0.066.$$

π_Q is the quality factor, and since a JANTX part is used $\pi_Q = 1.0$. π_E is the environmental factor, and for the missile launch application, $\pi_E = 32.0$.

All of this is brought together to compute the failure rate. The product of the adjustment factors is computed as

$$\lambda = \lambda_b \pi_T \pi_A \pi_R \pi_S \pi_Q \pi_E \frac{\text{failures}}{10^6 \text{ hours}}$$

$$= \left(\frac{7.4 \times 10^{-4} \text{ failures}}{10^6 \text{ hours}} \right)(1.73)(0.7)(0.84)(0.066)(1.0)(32.0)$$

$$= \frac{1.59 \times 10^{-3} \text{ failures}}{10^6 \text{ hours}}$$

This allows estimation of the MTTF and requested reliability

$$\text{MTTF} = \frac{1}{\lambda} = 6.3 \times 10^8 \text{ hours} = \underline{71{,}804 \text{ years}}.$$

$$R(20 \text{ years}) = \exp\left(-\frac{20 \text{ years}}{71{,}804 \text{ years}}\right) = 0.9997$$

$$= \underline{99.97\%}.$$

In conclusion, the BJT is estimated to be highly reliable in 20 years.

8.2.3 Thermal Management and Power Derating

One of the quantities computed in Example 8.4 was the junction temperature T_J, which was computed from the power dissipated in the device and a quantity known as thermal resistance, θ. It is important to understand this in more detail, since it affects the reliability of microelectronic devices. Furthermore, if the junction temperature exceeds a certain value, the device will fail. Thus, thermal management issues need to be taken into account. We start with a physical model in Figure 8.8(a), which has a junction (the integrated circuit or device), enclosed by a case (the packaging of the device), surrounded by ambient environmental conditions. In part (b) a heat sink is included, which aids in thermal transfer. Each element has associated with it a quantity known as thermal resistance that measures the ability of that particular element to transfer heat to another element. A result from heat transfer for electronics is that changes in temperature (ΔT) are proportional to the product of power dissipation (P_D) and the thermal resistance. This relationship is

$$\Delta T = P_D \theta. \tag{27}$$

It is similar to Ohm's law in that the change in temperature, power, and thermal resistance (units = °C/W) are analogous to voltage, current, and electrical resistance respectively. The total thermal resistance between two elements, such as between ambient and the junction, is the sum of all thermal resistances between them. In the case with no heat sink, this produces a junction-to-ambient resistance of $\theta_{JA} = \theta_{JC} + \theta_{CA}$, while in the case with a heat sink, $\theta_{JA} = \theta_{JC} + \theta_{CS} + \theta_{SA}$. In the case of the heat sink, θ_{CA} is replaced by the sum $\theta_{CS} + \theta_{SA}$, which has a lower combined thermal resistance and greater ability to dissipate heat.

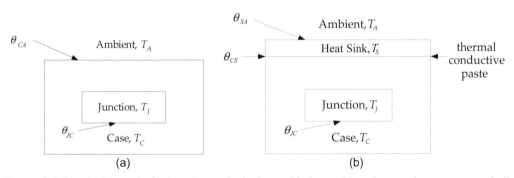

Figure 8.8 Physical model of microelectronic devices with thermal junctions and temperatures indicated. (a) Device inside casing. (b) Device inside casing with a heat sink added.

This Ohm's law type of relationship means that thermal transfer can be modeled by familiar resistive circuits, as shown in Figure 8.9. From this circuit model, the temperature can be found at different points from the thermal resistance and power dissipation. Most important, the junction temperature is found as

$$T_J = T_A + P_D \theta_{JA}. \tag{28}$$

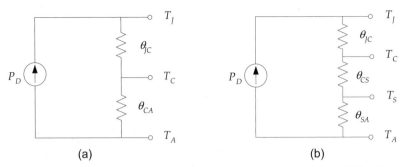

Figure 8.9 Resistive models for thermal transfer in microelectronic devices. (a) Device with no heat sink. (b) Device with a heat sink added.

Let's now apply these results. Manufacturer datasheets typically identify the absolute maximum power dissipation and note that the device should be derated if operated at ambient conditions above room temperature ($T_A = 25°C$). The data sheets also supply a maximum junction temperature for the device. It is clear from the resistive model that, for a fixed power dissipation, the junction temperature increases along with ambient temperature. If the maximum junction temperature is exceeded, the device will be destroyed. Another way to look at this is that as ambient temperature increases, the maximum amount of power a device can dissipate decreases. This decrease in maximum power dissipation is known as *derating*. From (28), the maximum power that can be dissipated in a device at a given ambient temperature is

$$P_{D,\max} = \frac{T_{J,\max} - T_A}{\theta_{JA}}. \tag{29}$$

From this relationship, a power derating curve is plotted in Figure 8.10, showing the maximum power versus ambient temperature. Example 8.5 demonstrates the application of this to the inverter in Example 8.4.

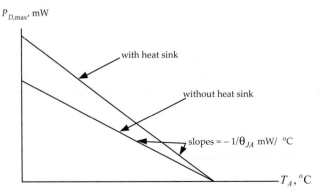

Figure 8.10 Typical power derating curves.

Example 8.5 Power Derating for the Inverter Circuit.

Problem: Assume the circuit in Example 8.4 is operating at an ambient temperature $T_A = 120°C$ and that no heat sink is used. (a) Determine the derated power and whether the design is within the manufacturer's limits for power dissipation at this temperature, and (b) recompute the reliability at 20 years at this elevated operating temperature.

Solution:
(a) From the manufacturer data sheet in Appendix D, the 2N3904 BJT has a thermal resistance of $\theta_{JA} = 200°$ C/W and a maximum junction temperature of 150°C. From this the maximum power is computed from (29) as

$$P_{D,\max} = \frac{(150 - 120)° \text{ C}}{200° \text{ C/W}} = 150 \text{ mW}.$$

The derated, or maximum, power at this temperature is 150 mW. Clearly, the 125 mW of power dissipated as determined in Example 8.4 for the BJT is within this derated limit.

(b) To compute the failure rate the junction temperature and π_T are recomputed.

$$T_J = T_A + P_D \theta_{JA} = 120°C + (125 \times 10^{-3} \text{ W}) \left(\frac{200°C}{W} \right) = 145°C$$

$$\pi_T = \exp\left[-2114 \left(\frac{1}{145+273} - \frac{1}{298} \right) \right] = 7.66.$$

With this new value, the value of $\lambda = 7.00 \times 10^{-3}/10^6$ hours, and the reliability is reduced slightly to 99.88%. Note, however, the junction temperature is quite high at 145°C and further increases in temperature would likely destroy the device.

8.2.4 Limits of Reliability Estimation

It must be kept in mind that the reliability estimates are just that, estimates, and there are limitations in their use. First, realize that the failure rate data comes from accelerated stress tests, where devices are put under stress beyond normal operating conditions, and from these the failure rates are estimated. (Nobody sits around waiting 20 years for the devices to fail!) The tests are based upon mathematical models for the failure rate and the device lifetime. Second, there are other factors that influence reliability that are not addressed by λ, such as the manufacturing processes used, the quality of manufacturing technologies, shock, and corrosion. Part of the value of reliability estimation is for comparative purposes in evaluating different design options. Applying these methods forces the designer to consider the operating conditions and factor them into the design.

8.3 System Reliability

The previous section focused on determining the reliability of a single device. It is natural to ask, "How can the reliability of a system consisting of many devices be determined?" In order to derive the overall reliability of a multicomponent system, it is necessary to take into account the overall system structure.

8.3.1 Series Systems

Consider the inverter circuit in Example 8.4—failure of any one component in the circuit would lead to the failure of the overall system or circuit. Conceptually, a system in which the failure of a single component (or subsystem) leads to failure of the overall system is known as a **series system**. Figure 8.11 shows a block diagram of a series system composed of boxes $S_1, S_2, \ldots, S_n$, that represent the components, or the subsystems, of a larger system.

Figure 8.11 A series system consisting of components, or subsystems $S_1, S_2, \ldots, S_n$.

To compute the overall reliability of a series system, $R_s(t)$, it is assumed that the failure of subsystems or components are independent events. The system is operable only if subsystems S_1 and $S_2 \ldots$ and S_n are all simultaneously operating. Therefore, the probability of the overall system operating is given by the product of reliabilities for all of the subsystems as follows

$$R_s(t) = R_1(t) R_2(t) \cdots R_n(t) = \prod_{i=1}^{n} R_i(t). \quad (30)$$

It is important to remember that failures are assumed to be independent events, just as flipping a coin twice is considered two independent events. The overall system reliability is less than or equal to that of any single subsystem, since all reliability values are ≤ 1. Thus R_s decreases as the number of subsystems increases. Assuming a constant failure rate for all system components gives the following result for the overall system reliability

$$R_s(t) = e^{-\lambda_1 t} e^{-\lambda_2 t} \cdots e^{-\lambda_n t} = \exp\left(-\sum_{i=1}^{n} \lambda_i t\right). \quad (31)$$

This leads to a series system failure rate and MTTF of

$$\lambda_s = \sum_{i=1}^{n} \lambda_i \quad \text{and} \quad \mathrm{MTTF}_s = \frac{1}{\lambda_s}. \quad (32)$$

Example 8.6 revisits the inverter problem where the failure rates of all components are considered for system reliability estimation.

Example 8.6 Inverter Circuit Reliability.

Problem: For the system in Example 8.4, estimate (a) the overall system reliability in 20 years, and (b) the MTTF. Assume room temperature and that ¼-watt fixed composition resistors are used.

Solution:

(a) Conceptually this is a series system—if any of the individual components fail, then the overall system will fail. That means that failure rates for the two resistors are needed in addition to the value previously computed for the transistor. They depend upon the power dissipated in each resistor, which is 125 mW and 0.9 mW for the collector and base resistors respectively. The failure rate for a fixed composition resistor, from MIL-HDBK-217F, is

$$\lambda_{resistor} = \lambda_b \pi_R \pi_Q \pi_E \text{ failures}/10^6 \text{ hours}$$

For the collector resistor, R_C the base failure rate is computed as

$$\lambda_b = 4.5 \times 10^{-9} \exp\left[12\left(\frac{T+273}{343}\right)\right] \exp\left[\frac{S}{0.6}\left(\frac{T+273}{273}\right)\right]$$

$$= 4.5 \times 10^{-9} \exp\left[12\left(\frac{25+273}{343}\right)\right] \exp\left[\frac{0.125/0.25}{0.6}\left(\frac{25+273}{273}\right)\right]$$

$$= 3.77 \times 10^{-4}.$$

The S term is the ratio of power dissipated to the maximum power rating. The values $\pi_R = 1.0$, $\pi_Q = 15.0$, and $\pi_E = 27.0$ are directly read from tables. Thus the overall failure rate for the collector resistor is

$$\lambda_{resistor1} = (3.77 \times 10^{-4})(1.0)(15.0)(27.0) = 1.53 \times 10^{-1} \text{ failures}/10^6 \text{ hours}.$$

The process for the base resistor R_B is similar, and results in

$$\lambda_{resistor2} = 6.1 \times 10^{-2} \text{ failures}/10^6 \text{ hours}.$$

The total failure rate is given from (32) as

$$\lambda_s = \lambda_{BJT} + \lambda_{resistor1} + \lambda_{resistor2} = 0.215 \text{ failures}/10^6 \text{ hours}$$

$$R_S(t) = \exp(-\lambda_s t) = \exp\left(-\frac{0.215}{10^6 \text{ hours}} \times \frac{24 \text{ hours}}{\text{day}} \times \frac{365 \text{ days}}{\text{year}} \times 20 \text{ years}\right) = 0.963$$

$$= \underline{96.3\%}.$$

Since resistors are pretty reliable devices, the overall system reliability decreases only a small amount relative to that of the BJT itself.

(b) The MTTF is given by $1/\lambda_s$, which in this case is <u>531 years</u>.

8.3.2 Parallel Systems

From (30) it is clear that as more components are added to a series system, the reliability decreases. It is natural to ask if the reliability can be increased. The use of redundancy gives us a method to answer in the affirmative. A design has **redundancy** if it contains multiple modules performing the same function where a single module would suffice. By its very nature redundancy allows improperly functioning modules to be switched out of the system without affecting its behavior. With redundancy the overall system functions correctly when any one of the submodules is functioning. Figure 8.12 shows a simplified view of a **parallel system** with subsystems $S_1, S_2, \ldots, S_n$.

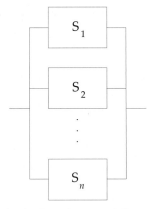

Figure 8.12 A parallel, or redundant, system consisting of subsystems $S_1, S_2, \ldots, S_n$

In order to compute the reliability of a parallel system, note that a parallel system functions correctly when S_1 is functioning correctly, or S_2 is functioning correctly, ... or S_n is functioning correctly. It would be nice if it were possible to write an equation stating that $R_s(t) = R_1(t) + R_2(t) + \cdots + R_n(t)$, where + is the logical OR operator. Unfortunately, there is no direct way to realize the OR operation in probability theory. This is resolved by working with failure function $F(t)$ instead. The probability that the system will fail by time t, $F_s(t)$, is equal to the probability that subsystem S_1 will fail and S_2 will fail and ... S_n will fail. This probability is expressed mathematically as

$$F_s(t) = F_1(t)F_2(t)\ldots F_n(t) = \prod_{i=1}^{n}F_i(t) = \prod_{i=1}^{n}[1-R_i(t)]. \tag{33}$$

The overall system reliability of the parallel system is found from this as

$$\boxed{R_S(t) = 1 - F_s(t) = 1 - \prod_{i=1}^{n}[1 - R_i(t)]}. \tag{34}$$

As more redundant components are added to a parallel system, additional $1 - R_i(t)$ terms are introduced into the product term. This decreases the value of the product, hence the overall system reliability is increased as more redundant systems are added.

In order for a parallel system to work, a mechanism must be in place to monitor each of the subsystems to make sure that they are operating correctly. Developing circuits to detect failures and control the switching between subsystems can be complex and is not considered here. Special care must be paid to the switching circuit itself, as malfunction of this circuit could lead to an overall system failure. An example of parallel system reliability is given in Example 8.7.

Example 8.7 Reliability of a Redundant Array of Independent Disks (RAID).

Problem: In a RAID, multiple hard drives are used to store the same data, thus achieving redundancy and increased reliability. One or more of the disks in the system can fail and the data can still be recovered. However, if all disks fail, then the data is lost. For this problem, assume that the individual disk drives have a failure rate of $\lambda = 10$ failures/10^6 hours. How many disks must the system have to achieve a reliability of 98% in 10 years?

Solution: The reliability of a parallel system with redundancy is given by (34). Since all of the disks are identical, the expression simplifies to

$$R_S(t) = 1 - [1 - R_i(t)]^n$$

$$0.98 \leq 1 - \left[1 - \exp\left(-\frac{10}{10^6 \text{ hours}} \times \frac{24 \text{ hours}}{\text{day}} \times \frac{365 \text{ days}}{\text{year}} \times 10 \text{ years}\right)\right]^n$$

$$0.98 \leq 1 - [1 - 0.42]^n$$

$$\Downarrow$$

$$0.02 \leq (0.58)^n$$

$$\Downarrow \qquad \text{(take logarithm of both sides)}$$

$$\log(0.02) \leq n \log(0.58)$$

$$\Downarrow$$

$$n \geq 7.2$$

In order to achieve this reliability, $\underline{n = 8}$ disks are required. The reliability of each individual disk is low at 42%, but with redundancy, the overall system reliability is quite high.

8.3.3 Combination Systems

Many real systems do not fit neatly into either parallel or series reliability models, as shown in Figure 8.13. Rather, they may be a combination of the two, and such systems will be referred to as combination systems. One way to determine the reliability of a combination system is to utilize the results obtained for series and parallel systems in (30) and (34). The system network is reduced by combining parallel subsystems into a single block, whose reliability is given by (34), while series subsystems are reduced to a single block whose reliability is given by (30). This is conceptually analogous to combining series and parallel resistances in electrical circuits. The network is continually reduced until only a single block remains whose reliability is known from all of the subsystem combinations.

To illustrate this, consider the system in Figure 8.13. To determine the reliability, start by combining the three parallel systems S_2, S_3, and S_4, whose combined reliability is determined by application of (34) to be $R_{s_{2,4}}(t)=1-(1-R_2(t))(1-R_3(t))(1-R_4(t))$. The result is then combined with S_1 in series to give the overall system reliability, $R_s(t) = R_1(t)[1-(1-R_2(t))(1-R_3(t))(1-R_4(t))]$. The chapter concludes with Example 8.8, which addresses combination system reliability.

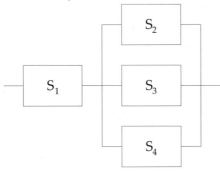

Figure 8.13 A combination series-parallel system. S_2, S_3, and S_4 are redundant parallel systems.

Example 8.8 Combination System Reliability.

Problem: Consider the system shown below with the following reliabilities at a fixed time t, $R_1 = R_2 = 80\%$. Determine the reliability that subsystems R_3 and R_4 must have so that the overall system reliability is greater than 95%.

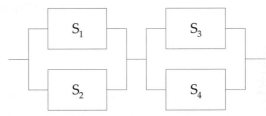

Solution: The parallel systems can be combined into single systems whose reliabilities are
$$R_{s_{1,2}} = 1 - (1 - R_1)(1 - R_2)$$
$$R_{s_{3,4}} = 1 - (1 - R_3)(1 - R_4).$$
They are combined in series to give the overall system reliability
$$R_s = \left[1 - (1 - R_1)(1 - R_2)\right] \times \left[1 - (1 - R_3)(1 - R_4)\right].$$
Substituting values and assuming $R_3 = R_4$ gives
$$0.95 = \left[1 - 0.2^2\right] \times \left[1 - (1 - R_{3,4})^2\right].$$

Solving for the reliabilities gives the final result
$$R_3 = R_4 = 0.90.$$
This example demonstrates the power of redundant systems. S_1 and S_2 have somewhat low reliabilities relative to the overall system goal, but the reliability of the parallel combination of S_1 and S_2 is 96%. It requires a reliability for systems 3 and 4 of $R_3 = R_4 = \underline{90\%}$ while the combined reliability of systems 3 and 4 is 99%.

8.4 Summary and Further Reading

This chapter presented the basics of probability theory and methods for estimating the reliability of components and systems. Failure rate is an important quantity that is determined empirically and provides the rate of failure over the lifetime of a component or system. A mathematical definition of reliability was derived from this quantity, which takes a simple exponential form in the case of a constant failure rate. This was applied to estimate the reliability of single components, particularly using failure rates from MIL-HDBK-217F. Issues of thermal transfer and power derating were considered. Reliability estimation was extended to more realistic systems consisting of multiple components in series and parallel forms. The use of redundancy with parallel systems to increase the overall system reliability was addressed.

There are plenty of good textbooks available on the probability theory, if it is necessary to study probability theory further. The book *Practical Reliability of Electronic Equipment and Products* [Hna03] provides detailed coverage for electrical systems reliability. It includes factors not considered here such as thermal management on printed circuit boards, procurement practices, and electromagnetic interference. Two excellent articles that demonstrate the application of design for reliability and redundancy for an embedded system application are by George Novacek in *Circuit Cellar* magazine [Nov00, Nov01].

8.5 Problems

8.1 Consider a random variable that obeys a uniform density and varies from 2 to 5. (a) Determine the mean and variance of the random variable. (b) What is the probability that the random variable is between 2 and 3? (c) Plot the CDF.

8.2 In Figure 8.7 it was assumed that the CDF function is monotonically increasing. That is, $F(t) < F(t+\Delta t)$. Show why this is so, using equation (13).

8.3 Describe what is meant by *failure rate, failure function,* and *reliability*.

8.4 Consider an integrated circuit that has $\lambda = 50/10^6$ hours. (a) Determine the mean time to failure. (b) Determine the reliability in 5, 10, 15, and 20 years.

8.5 Consider a CD4001BC two-input quad NOR Gate (data sheet available in Appendix D). Also assume it is a glass-sealed dual-in-line package (DIP), 15 years in production,

$\theta_{JA} = 70°$ C/W, the power dissipated in the application averages 10 mW, it has B-1 quality, and it is to be operated in laboratory equipment at an ambient temperature of 25°C. Use the MIL-HDBK-217F data in Appendix C to determine the MTTF and estimate its reliability in 25 years.

8.6 Use the MIL-HDBK-217F data in Appendix C to estimate the reliability of a 32-bit CMOS microprocessor. Assume that it is used in a missile launcher, the ambient temperature is 120°C, that it has 64 pins, that it has been in production for 6 months, B-1 quality parts are used, and it is a nonhermetic DIP. Determine the MTTF and reliability for the microprocessor in 20 years. (Note: this is the same operating environment used for the BJT in Example 8.5.)

8.7 Consider a 1N4001 diode (data sheet in Appendix D) that is to be operated in the switching circuit shown below. Also assume that the part quality is lower, it is metallurgically bonded, and that it is to be used in an airborne inhabited cargo environment at an ambient temperature of 50°C. Use the MIL-HDBK-217F data in Appendix C to determine the MTTF and estimate its reliability in 25 years.

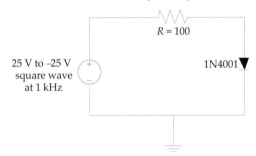

8.8 Use the MIL-HDBK-217F data in Appendix C to determine the reliability of the inverting op amp circuit, shown below, in 15 years. Assume that it is used in an automotive application (environmental factor = G_M), the ambient operating temperature is 80°C, industrial quality parts are employed, and ¼-watt fixed composition resistors of the lowest quality are used. The data sheet for the LM741 op amp is in Appendix D. The LM741 is considered to be a linear microcircuit, comes in a dual-in-line package (DIP), contains 25 bipolar transistors, has S quality, and has been in production for well over 20 years.

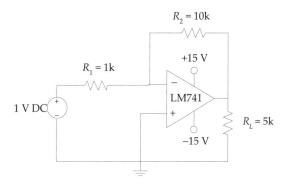

8.9 Consider the circuit in Example 8.4. Assume that a heat sink is attached to the 2N3904 BJT (data sheet in Appendix D) and it has the following thermal resistance values and $\theta_{CS} = 10°C/W$ and $\theta_{SA} = 8°C/W$. If the device is operated at an ambient temperature of 130°C, determine the maximum power dissipation of the BJT and its reliability in 50 years.

8.10 Your company intends to design, manufacture, and market a new RAID (redundant array of independent disks) for network servers. The system must be able to store a total of 500 GB of user data and must have a reliability of at least 95% in 10 years. In order to develop the RAID system, 20 GB drives will be designed and utilized. To meet the requirement, you have decided to use a bank of 25 disks (25 × 20 GB = 500 GB) and utilize a system redundancy of 4 (each of the 25 disks has a redundancy of 4). What must the reliability of the 20 GB drive be in 10 years in order to meet the overall system reliability requirement?

8.11 S_1 has a failure probability of 2% and S_3 has a failure probability of 3% $S_{2,1}$, $S_{2,2}$, $S_{2,3}$, and $S_{2,4}$ are identical redundant systems. Determine the required reliability of the redundant systems necessary for the overall system to have a reliability of 94%.

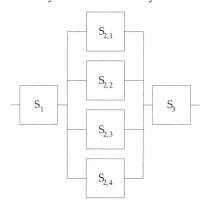

8.12 Consider the design of a triply redundant majority voting system, with three binary inputs a, b, and c shown below. The inputs represent data from three independent sources from which the objective is to determine if the majority of the input bit values are logic level 0 or 1. Each majority circuit outputs the bit value that is in the majority of the inputs. The output of each majority circuit is fed into a resistor and LED (light-emitting diode) network. The LEDs are lit if the output of the majority circuit is a logic 1, otherwise the LED is off. Ideally, all three LEDs are lit if a majority of inputs are 1, else they are all off. However, if part of the system fails, the LED readings may not be reliable, so a majority-rules decision is used on the LEDs. The criterion used is that if two or more LEDs are on, then a majority of inputs are considered to be high, otherwise, a majority of inputs are considered to be low. Determine the probability of a false reading based upon this criterion if each component in the system (gate, resistor, and LED) has a reliability of 90%.

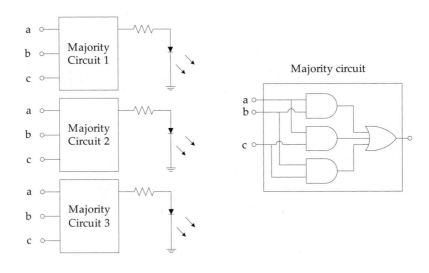

Part III — Professional Skills

Chapter 9 Teams and Teamwork

The way a team plays as a whole determines its success. You may have the greatest bunch of individual stars in the world, but if they don't play together, the club won't be worth a dime.— Babe Ruth

Engineering projects are often very large and require too many diverse skills to be completed by a single person. This necessitates the use of teams, and it is likely that you will participate in many during your professional career. That is why employers of engineering graduates consistently rate teamwork ability as one of the most desirable attributes in a potential employee. Teams can collectively perform tasks that could not possibly be completed by individuals working in isolation. However, just because a team is created does not mean that it will function effectively, and examples of failed teams abound. When people learn that they are going to work on a team, their initial reaction is often one of dread, usually due to bad experiences in the past. The good news is that with some attention to team principles and a conscious effort, working on a team can be a positive experience. Much research has been done and much has been written on teams and teamwork. The objective of this chapter is to provide a primer on what constitutes a team, review present models of team development, identify the characteristics of effective teams, and provide guidance for effective teamwork.

Learning Objectives

By the end of this chapter, the reader should:

- Understand the characteristics that define a team and understand why a team is formed.
- Understand different models for the stages of team development.
- Understand the characteristics of effective teams.
- Be able to develop team process guidelines.

9.1 What Is a Team?

The concept of a team is used broadly in many different contexts, so the following definition of a team proposed by Katzenbach and Smith in <u>The Wisdom of Teams</u> [Kat93] is provided for the basis of our discussion:

> *A small group of people with complementary skills, who are committed to a common performance, performance goals, and approach for which they hold themselves mutually accountable.*

This concept of a team is highly relevant to engineering design projects.

The definition indicates that a team should be a small group of people, typically two to ten. Teams larger than this become very difficult to manage. The reason is that the number of relationships between people increases rapidly with an increase in the number of people on the team. The number of person-to-person relationships is equal to $n(n-1)/2$, where n is the number of people on the team. Any one of these relationships can falter, causing problems. In addition, it is hard to develop consensus on important issues as the number of people on the team increases.

Second, the definition indicates that teams should be composed of members who have complementary skills. Members should be selected for the background and skills that each person brings to the team, not for personality and friendship. This leads to the related concepts of *cross-functional* and *multi-disciplinary teams.* Cross-functional teams are those that are composed of people from different organizational functions, such as engineering, marketing, and manufacturing [Wil95]. Cross-functional teams are particularly important in new product development, where multiple functions are required to bring a product from concept, to manufacture, and ultimately to the market. Multidisciplinary implies that the team is composed of members from different disciplines. There is no agreement in the engineering community as to what exactly constitutes a multidisciplinary team. In general, the idea of complementary skills applies where a team may have representation from multiple technical disciplines. For example, the development of a robot would require members with multidisciplinary expertise in computing, electronics, and mechanical systems.

Third, the definition states that teams should have common performance goals. The goals take time to develop and the team will likely flounder in creating them, but without specific goals to achieve there is no need for a team. In the context of design, the team's goals are defined by the problem statement and requirements specification developed in Chapters 2 and 3 respectively. They collectively state what the team is trying to achieve and provide a way for the team to verify its success—are the requirements met and can the team demonstrate that they are?

Finally, the team definition indicates that teams should have a common approach for which they hold themselves mutually accountable. This can be a common approach to solving problems, such as application of the design process. It can also be expressed in terms of a method for holding the team members accountable. One of the biggest problems that teams face is handling situations where the team fails to meet a given objective. This may be due to one or more members failing to meet their commitments. The team must be able to hold each other mutually accountable in a constructive manner in these situations.

9.2 Models of Team Development

Bruce Tuckman developed a model for team development that contains fives stages, known as forming, storming, norming, performing, and adjourning [Tuc65]. It is instructive to have an understanding of these stages so that you are aware of the dynamics that occur as a team develops. The five stages are described as follows:

- *Forming.* This is the stage where the team is created. The team members may not know each other, leading them to be anxious and uncomfortable. It is likely that the team's objectives are not well defined and that the members' roles are ambiguous.

- *Storming.* In this stage, the team works to develop its objectives, while the members try to define their roles. Conflict often appears as members jockey for position and advantage. It is characterized by fighting and power struggles, which may be overt, but in many cases are subtle and below the surface. The team must resolve individual versus group goals. For example, what if the team is trying to select a project concept? Some members may have strong opinions and preconceived notions about what the project should be. Failure to navigate the storming stage means that the team will not reach the performing stage and will be forced to return to the storming stage at some time.

- *Norming.* In this stage the team starts to become more cohesive. Members accept the team's objectives and their roles on the team, and they agree upon procedures. In this phase the team develops common approaches for solving problems and managing conflicts.

- *Performing.* In this stage, the team focuses on performing tasks and achieving the objectives set forth. The team should be collaborating and easily making decisions. Disagreements that arise are accepted and resolved by the team norms.

- *Adjourning.* Here, the team dissolves, hopefully as a result of successful project completion. However, dissolution can occur abruptly if the project is cancelled, or even worse, if the team is unable to function together.

Katzenbach and Smith proposed the team performance curve in Figure 9.1, which shows the performance impact of teams versus varying levels of effectiveness. The types of teams they defined are the working group, the pseudoteam, the potential team, and the high-performance team. The most closely related stages of team development from the Tuckman model are indicated in parentheses on the figure. Notice that the scale on the x axis is team effectiveness, not time. There may be an implicit assumption that teams will move through these stages as a function of the time the team is together. That is certainly a desired goal, but not necessarily what the model implies.

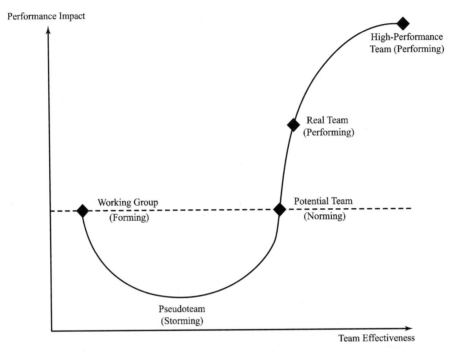

Figure 9.1 The team performance curve proposed by Katzenbach and Smith [Kat93]. This shows the performance impact of teams versus varying levels of effectiveness. The Tuckman stages of team development (forming, storming, norming, performing) that are most closely related to the performance points are superimposed on the model [Tuc65].

The *working group* is defined as a group of individuals working in isolation, who come together occasionally to share information. In other words, if every member of the team works in isolation and meets only to share ideas, they would achieve the performance of the working group. This level of performance is represented by the dashed baseline. In fact, the working group is not a team as given by the definition and serves a very different purpose. The *pseudoteam* represents an under-performing team where the sum effort of the team is below that of the baseline performance, while the *potential team* is one in which the team is functioning at a level equal to that of the working group. At a minimum, teams should at some point perform above the level of a potential team, otherwise, there is no need for a team. The objective is to perform at the level of a *real team*, where the performance exceeds that of the working group. In rare instances the level of a *high-performance team* may be realized in which the team significantly outperforms all similar teams.

9.3 Characteristics of Real Teams

How can a team reach the performing stage and simultaneously become a real team? Unfortunately, there is no set process for guaranteeing that a team will become a real team. There are characteristics and principles of effective teams, and it has been observed that real teams apply these principles. However, realize that there is a difference between a team and teamwork principles. Real teams apply teamwork principles but this does not imply that the converse statement is true—that applying these principles will result in a real team. The lesson is that successful teams adhere to good teamwork principles, but the application of good teamwork principles does not automatically guarantee that teams will be successful. However, ignoring teamwork principles almost always leads to failure. The remainder of this section presents some of the characteristics of real teams.

9.3.1 Select Members for Their Skills

Selecting team members for their skills is a key to success identified by Katzenbach and Smith. They defined three categories of relevant skills: (1) technical and functional, (2) problem solving, and (3) interpersonal. The right mix of technical skills on a design project is important to achieve the desired objectives, reinforcing the need for multidisciplinary and cross-functional teams. Different problem-solving approaches and the mix of interpersonal skills are important as well, but harder to determine. The Meyers-Briggs personality tests or the Keirsey temperament sorter [Kei84] are tools that can be used to identify different problem-solving approaches and personality traits, but in themselves are not the answer for selecting teams.

One important point to consider is whether teams should be formed by self-selection or assigned by a supervisor. There are pro and con arguments for each method of selection. An argument for self-selection is that members believe the objectives of the team are important, leading to a higher level of commitment. A potential pitfall is that not enough attention is paid to the skills necessary to complete the project. In the second case, teams are assigned by someone who has an understanding of the skills needed for the project and assigns members accordingly. A potential pitfall is that this may create animosity from team members who are dissatisfied with the project from the outset.

9.3.2 Identify Objectives

Teams are created to realize shared goals, and if the goals are not well defined, the motivation for being part of the team is not clear. In the context of engineering design, the combined problem statement and requirements specification presented in Chapters 2 and 3 serve that purpose. The problem statement describes what the team is trying to achieve, while the requirements specification sets verifiable targets that define success. Both items are so important that there should be a consensus agreement among the team members on their content. The team should be challenged by targets that are aggressive, yet achievable. Identifying measurable objectives was identified by Katzenbach and Smith as one of the most important attributes of successful teams. Further, all team members must be committed to achieving the objectives.

9.3.3 Develop Decision-Making Guidelines

Teams regularly make decisions that affect all of its members and the success of the project. The team must determine how decisions are to be made, and once they are, all members need to accept the outcome. The team also needs to understand the importance of different decisions—not all are equally important, and different approaches should be used, depending upon the impact of the decision. Models for decision making are outlined as follows [Joh02]:

1. *Decision by Authority*. The leader makes all decisions, typically without discussion by the rest of the team. This is effective for fast decision making, but often does not lead to the best decision. It can also produce resentment among the team.

2. *Expert Member*. The most expert member on the subject decides. This is effective in cases where there is a single member who is clearly the expert on the subject.

3. *Average Member Opinion*. The average team member opinion is used. A method needs to be devised to determine what constitutes the average opinion.

4. *Decision by Authority after Discussion*. The leader makes a decision after all team members discuss the issue and provide input.

5. *Minority Control*. A few members act as a subcommittee and solve the problem. This makes sense if the minority group consists of experts on the particular problem.

6. *Majority Control*. A simple majority is used to make decisions.

7. *Consensus*. All team members must agree to and commit to the decision. This generally comes after much discussion and evaluation of the different alternatives. This is the best approach, and the most time-consuming. It is not necessary for all decisions to be made this way, but consensus should be reached for important decisions.

A formal and widely used method for brainstorming and team decision making is known as the **nominal group technique** (NGT) [Del71]. NGT was covered in Chapter 4 and is repeated here. In NGT, each team member silently generates solutions to a problem and the ideas are then reported out in a round-robin fashion until all ideas are exhausted. Then, each member gets to cast a predetermined number of votes for the ideas presented. The top idea is then selected, or alternatively, the top few ideas are discussed further and voted upon again. The steps of NGT are outlined below:

1. *Read problem statement*. It should be read out loud by a team member (the facilitator).

2. *Restate the problem*. Each person restates the problem in his or her own words to ensure that all members understand it.

3. *Silently generate ideas.* All members silently generate ideas during a set period of time, typically 5–15 minutes.

4. *Collect ideas in a round-robin fashion.* Each person presents one idea in turn until all ideas are exhausted. The facilitator should clarify ideas and all should be written where the entire team can view them.

5. *Summarize and rephrase ideas.* Once the ideas are collected, the facilitator leads a discussion to clarify and rephrase the ideas. This ensures that the entire group is familiar with them. Related ideas can be grouped or merged together.

6. *Vote.* Each person casts a predetermined number of votes, typically three to six, for the ideas presented. The outcome is a set of prioritized ideas that the team can further discuss and pursue.

9.3.4 Hold Effective Meetings

Teams need to meet for a variety of reasons, such as determining objectives, tracking progress, assigning tasks, preparing deliverables, and resolving problems. It is in these meetings that teamwork principles are most critical and where problems can arise. In the workplace it is estimated that people spend half their time in tasks that are related to meetings (preparing, attending, and following up). Three elements of effective meetings are: (1) well-defined roles for the participants, (2) a structure for conducting the meeting, and (3) the application of interpersonal skills [Bel94]. Poorly organized and ineffective meetings lead to cynicism and poor team performance. The structure of meetings does not always need to be the same. Some keys to effective meetings are:

- *Have an agenda.* Agree upon the goals for a meeting in advance.

- *Show up prepared.* All members should show up on time with their tasks completed.

- *Pay attention.* Each person should speak in turn, and there should be no disruptive side conversations. No person should monopolize the conversation and all points of view should be heard.

- *Agree upon a meeting time and place.* People have busy schedules and different work habits. Believe it or not, failure to agree upon this basic issue can lead to tremendous conflict.

- *Summarize.* At the end of the meeting summarize what was discussed, important decisions made, and actions to be taken.

9.3.5 Develop Team Roles

As the group finds its norms, members should settle into different roles. They may be based upon the technical aspects of the design, such as hardware, software, and mechanical systems. The team may find that certain members are more effective at different tasks such as procuring parts, making presentations, and writing technical documents. Roles often evolve and change as the team develops.

Team roles are usually not known early on in the forming stage and members can be assigned formal roles to perform in at the outset. This is also common in group problem solving sessions. A typical set of roles is [App01]:

- *Leader.* The leader is not necessarily the authority figure, but should guide the team through the processes of decision making and problem solving. Effective leaders should ensure that all opinions are heard and that everyone is involved. Leaders coordinate tasks such as identifying the agenda.
- *Recorder.* The recorder is responsible for maintaining a record of the team's work, documenting important results, and recording responsibilities for different tasks.
- *Spokesperson.* The spokesperson articulates the team's results. The recorder and spokesperson roles are sometimes combined.
- *Optimist.* The optimist's role is to examine reasons why ideas and concepts will work, find their merit, and advocate other's ideas.
- *Pessimist.* The pessimist should challenge ideas and assumptions, and make sure that opposing ideas are presented and discussed.
- *Analyst.* The team analyst performs the important role of observing the team processes and provides feedback to the team for improvement.

The roles presented here are by no means definitive and other models having different roles can be utilized. The common thread is to have participants gain experience in the different roles so that they can become more effective team members. No matter what role a member is playing, they should all contribute to the solution of the problem at hand. As the team evolves and becomes more experienced, the need for formal roles will diminish.

9.3.6 Assign Tasks and Responsibilities

Nothing creates problems more than when team members do not have clearly defined responsibilities and tasks. The workload needs to be distributed in a fair manner and all members must perform real work. Without this, one or two people do most of the work, or even worse, nothing gets done. Chapter 10 presents project management principles, one of the primary aims of which is to develop and assign tasks to team members. Teams may consider designating one person as the project manager who is responsible for tracking the team's progress and ensuring that members are meeting their commitments. It is important that the project manager also be assigned tasks, outside of project management, that contribute to the project.

9.3.7 Spend a Lot of Time Together

Teams that spend a lot of time together are generally more successful [Kat93]. This is not only time together in meetings and working on project deliverables, but also includes extracurricular activities. All members of the team should be included, not just a subgroup. Experience shows that on teams of three students, two members may spend a great deal of time together and form a bond. This can alienate the third person (not necessarily intentionally) and lead to conflict.

9.3.8 Respect Each Other

Imagine a scenario where you are assigned to a team and one of the members is that quiet, weird guy who never says anything, or that loudmouth who never shuts up. You may have a real challenge ahead of you, but working with people who are not our best friends is a reality of life. You have to learn how to deal with personality issues if you want to be effective on teams. One key is to demonstrate respect for the other members of your team. Ways to do this are to:

- *Listen actively.* Most people are not effective listeners. As they listen to others they are mentally formulating their response, while not actually listening to the other person. By listening to the other person, then formulating an appropriate response, you will develop a better understanding of the other person's viewpoint and demonstrate that their opinion is being heard.

- *Consider how you respond to others.* How you respond to others affects your effectiveness. Are you negatively evaluating others' ideas or treating their ideas unfairly?

- *Constructively criticize ideas, not people.* It is fine to examine problems and constructively criticize ideas. Consider looking for ways to improve upon ideas that have flaws.

- *Respect those not present.* Nothing creates divisions and factions in a team faster than personally criticizing members not present. If a team member is not performing, refer to the team process guidelines examined in Section 9.4 for holding them accountable.

- *Communicate your ideas.* If you have an idea, clearly state it and be prepared to explain what the merits are. Be prepared for critical analysis and discussion of the idea.

9.3.9 Manage Conflicts Constructively

Conflicts inevitably occur on teams, and the more constructively they are addressed the better. Real teams do encounter conflicts, but they are adept at resolving them. Conflicts can be good, particularly when they lead the team to the solution of a problem or the development of consensus. They also provide an opportunity for members to express their opinions on important issues. Left unresolved and unspoken, they lead to resentment and suspiciousness, and escalate into personal conflicts. Conflicts may be false (you misinterpret another's behavior), based on performance (members aren't doing their work, or it is of poor quality), or based on procedure (disagreement with the way that meetings are run or decisions are made) [Dom01]. Some strategies for resolving conflicts are:

- *Focus on performance and ideas.* Focus on the performance of the team and not individual personality. Also, focus on ideas and constructively criticize them.
- *Listen to others.* It is very important to remain calm and listen carefully to others when conflicts arise.
- *Identify concerns.* If you have concerns about something, it is best to identify and address them, rather than hide them.
- *Apply the team's process guidelines.* In the next section, a format for process guidelines that govern the behavior and processes of a team is given. This should address a team's approach for resolving conflicts. It is important that the team must remember to apply the guidelines when conflicts arise.
- *Develop a plan to resolve the conflict.* Again, conflict can be positive and lead to the solution of problems. The team may not be able to resolve a conflict immediately, but can develop a plan for solving it within a specified period of time.
- *Mediation.* A mediator can be used after all avenues for resolution of the conflict have been exhausted. One way to do this is to employ a variation of the delphi technique. The delphi technique was originally developed as a brainstorming method to generate ideas anonymously from experts outside of a team or organization. It is applied for mediation through the following steps:
    1. Each member of the team should anonymously supply a description of the conflict and suggested remedies to the mediator.
    2. The mediator proposes a solution to the conflict, which is fed back to the team.
    3. The team members are given an opportunity to suggest modifications to the proposed resolution.
    4. Steps 1–3 are repeated until consensus is achieved.

9.4 Project Application: Team Process Guidelines

Teams should develop clear and detailed guidelines that govern their processes. These guidelines are based on the items covered in the previous section. The entire collection of guidelines forms the **team process guidelines**. Issues addressed in the team process guidelines include:

- *The team's name.* This creates an early opportunity for decision making.

- *The team's mission and objectives.* This can be a brief restatement of the design problem statement.

- *Decision-making guidelines.* Indicate what techniques are going to be used to make decisions and when they will be applied.

- *Meeting guidelines.* How meetings will run and what the expectations are for team members in the meetings.

- *Team roles.* If teams self-select, they should justify their team choice with identification of the complementary skills each member brings to the team.

- *Conflict resolution.* Identify how the team will resolve conflicts. For example, what will the team do when members are doing substandard work and not meeting the performance objectives? How will the team determine what substandard performance is? What happens if a member misses a meeting? What will the team do if they cannot resolve the conflict among themselves?

This document should be developed early on, referred to regularly throughout the project, and updated by the team as necessary. Table 9.1 contains a checklist and self-assessment of the team formation stage and processes.

9.5 Summary and Further Reading

This chapter touched on some of the fundamental concepts of team development that have been reported in the literature. Two well-known models of team development were presented. The first is the Tuckman model of the forming, storming, norming, performing, and adjourning stages. It is important for teams to carefully navigate the formative stages, agree upon the objectives, and reach the performing stage. The second is the Katzenbach and Smith team performance curve presented in Figure 9.1. Selected characteristics of real teams were identified. Finally, team process guidelines are a tool for identifying a team's norms, problem solving approach, and self-governing principles. A format for the guidelines was presented.

There are many excellent books and resources available that examine teams and teamwork. One that influenced this chapter greatly is The Wisdom of Teams by Katzenbach and Smith [Kat93]. It is a seminal work in the field and a widely recognized resource that addresses how to create effective teams. The forming, storming, norming, performing, and adjourning model of team development was proposed by Bruce Tuckman [Tuc65] and is widely documented as a good model for team development. The Team Training Workbook [Bel94] is another valuable resource developed by Arizona State University engineering faculty members. Teamwork and Project Management by Karl Smith [Smi04] is a short introduction to both teamwork principles and project management for engineering students. Free personality tests and temperament sorters can be found online, one example being the Keirsey temperament sorter at *www.advisorteam.com*.

Table 9.1 Checklist and self-assessment for team formation and processes (1 = strongly disagree, 2 = disagree, 3 = neutral, 4 = agree, 5 = strongly agree).

Team Formation	Score
The team's objectives are clearly defined.	
There is consensus among all team members that the objectives are the correct ones.	
The team members' complementary skills (technical, functional, interpersonal) have been identified.	
There are enough members on the team to cover all of the necessary competencies.	
There are not too many members on the team.	
Team Processes	
The team has developed clear guidelines for resolving conflicts and disagreements.	
The team has developed effective guidelines for holding all members of the team mutually accountable for achieving the objectives.	
The team has developed a strategy for holding effective meetings.	
The team has agreed upon a mutual meeting time and place.	
The team members trust each other.	
The team members demonstrate respect for each others ideas.	

9.6 Problems

9.1 Explain the difference between cross-functional and multidisciplinary teams.

9.2 Identify the characteristics of the forming, storming, norming, and performing stages in team development.

9.3 Describe the distinction between teams and teamwork.

9.4 According to this chapter, it is difficult to develop a consensus as the number of team members increases. Consider the situation where a team needs to agree on a proposal. Furthermore, assume that each team member's vote is random, with a 50% chance of agreeing with the proposal. Plot the probability of the team unanimously agreeing to the proposal versus the number of team members. Consider team sizes from 2 to 10. Overlay three additional plots for the situation where each team member has a 75%, 90%, and 99% chance of agreeing to the proposal.

9.5 **Project Application.** Develop team process guidelines as proposed in Section 9.4.

9.6 **Project Application.** Complete the team self-assessment in Table 9.1.

Chapter 10 Project Management

If you fail to plan, then you plan to fail. —Anonymous

The engineering community has led in the development of project management practices because building complex systems is a tremendous technical and managerial challenge. Currently, businesses tend to organize around projects that have significant value to the organization. Consequently, project management is consistently rated by employers as one of the most desirable skills sought in new college engineering hires [Par03]. The project management field includes topics such as initiating a project, team management, cost management, risk management, controlling, resource management, and performance management, to name a few. Many of these are addressed throughout this book from an engineering design viewpoint such as controlling (design process), initiating (project selection), performance management (requirements and testing), and team management.

The three important objectives of project management are to complete projects that are on time, within budget, and meet the requirements of the user. Since user requirements were addressed in Chapters 2 and 3, this chapter addresses the remaining two objectives of time and cost management. Time management introduces the work breakdown structure, which identifies the activities (combined tasks and deliverables) required to complete the project. Responsibility for completing the activities is then assigned to members of the team. Two graphical representations of the work breakdown structure, the network diagram and the Gantt chart, are introduced. These visual depictions show dependencies between tasks and allow for a quantitative analysis of the project plan. The chapter concludes with methods of cost estimation.

Learning Objectives

By the end of this chapter, the reader should:
- Be able to create a work breakdown structure.
- Be able to create network diagrams and Gantt charts.
- Be able to determine the critical path for completing a project and the float time for each activity in the plan.
- Be able to conduct break-even analysis and understand some basic methods of cost estimation.

10.1 The Work Breakdown Structure

The *work breakdown structure* (WBS) is a hierarchical breakdown of the tasks and deliverables that need to be completed in order to accomplish the project objectives. Creating the WBS is typically the first step in project planning. A WBS is an ordered set of activities required to complete the project. An *activity* is a combination of a task and its associated deliverables. *Tasks* are actions that accomplish a job, while *deliverables* are entities that are delivered to the project upon completion of tasks. Examples of deliverables are a circuit design, a software module, the integration and test of modules, a report, a presentation, and an approval. Example tasks include conducting research or writing a program.

The concept of the WBS was formalized by the United States Military in the 1993 document Work Breakdown Structure Handbook [MIL-HDBK 881]. The WBS has gained wide acceptance in industry and is described in MIL-HDBK 881 as follows:

- *A product-oriented family tree composed of hardware, software, services, data, and facilities. The family tree results from systems engineering efforts.*
- *A WBS displays and defines the product, or products, to be developed and/or produced. It relates the elements of work to be accomplished to each other and to the end product.*
- *A WBS can be expressed down to any level of interest. However the top three levels are as far as any program or contract need go unless the items identified are high cost or high risk. Then, and only then, is it important to take the work breakdown structure to a lower level of definition.*

This description indicates that WBS results from systems engineering efforts and that the structure of the WBS follows the design hierarchy. The second bulleted item focuses on the activities for the project. Gray and Larson [Gra02] recommend identifying the following activity attributes for each activity:

- A definition of the work to be done or delivered.
- A timeframe for completion of the activity.
- Resources needed to complete the activity.
- Person(s) responsible for the activity.
- Predecessors (or dependencies) for the activity. Predecessors are other activities that must be completed before the work can start.
- Checkpoints for monitoring progress.

The collection of activities and their attributes are gathered in the WBS table. The rows of the WBS table represent the project activities. These activities are arranged in a hierarchical fashion; each major activity is followed by its constituent subactivities. The columns of the WBS table are the activity attributes proposed by Gray and Larson. Table 10.1 contains the WBS table for the temperature monitoring design examined in Chapter 5 (Section 5.7).

Table 10.1 Example work breakdown structure for the design of a temperature monitoring system.

ID	Activity	Description	Deliverables / Checkpoints	Duration (days)	People	Resources	Predecessors
1	Interface Circuitry						
1.1	Design Circuitry	Complete the detailed design and verify it in simulation.	• Circuit schematic • Simulation verification	14	Rob (1) Jana (1)	• PC • SPICE simulator	
1.2	Purchase Components		• Identify parts • Place order • Receive parts	10	Rob		1.1
1.3	Construct and Test Circuits	Build and test.					
1.3.1	Current Driver Circuitry	Test of circuit with sensing device.	• Test data • Measurement of linearity	2	Jana (1) Rob (2)	• Test bench • Thermometer	1.2
1.3.2	Level Offset and Gain Circuitry	Test of circuit with voltage inputs.	• Test data • Measurement of linearity	3	Rob (1) Jana (2)	• Test bench	1.2
1.3.3	Integrate Components	Integrate the current driver and offset circuits.	• Test data verifying functionality and linearity requirement	5	Rob (1) Jana (1)	• Test bench • Thermometer	1.3.1 1.3.2
2	LED and Driver Circuitry						
2.1	Research A/D Converters	Make selection of A/D converter.	• Identify types, cost, and performance • Identify two potential converters for purchase	1	Alex	• Internet	
2.2	Complete Hardware Design	Design conversion hardware.	•Circuit schematic • Simulation verification	7	Ryan (1) Alex (2)	• Digital circuit simulator	2.1
2.3	Purchase LED and Driver Components		• Identify parts • Place order • Receive parts	10	Rob		2.2
2.4	Construct and Test	Test with supply voltage input.	• Test data showing digital output vs. voltage inputs	5	Alex (1) Ryan (2)	• Test bench • Logic analyzer	2.3
3	System Integration and Test	Complete integration of front-end and LED driver circuitry.	• Test data demonstrating functionality from temp input to LED output • System linearity measurement	7	Alex (1) Rob (1) Jana (1) Ryan (1)	• Test bench • Digital logic analyzer • Thermometer	1.3.3 2.4 (or 1 and 2)

Each activity is assigned an identification number (ID) and a name. The hierarchical nature of the activities is reflected in the ID numbering scheme. The three highest-level activities in the project plan are the Interface Circuitry (1), LED and Driver Circuitry (2), and System Integration and Test (3). The first two activities are further refined into subactivities shown by the indented items. For example, the Interface Circuitry activity contains three subactivities: Circuit Design (1.1), Purchase Components (1.2), and Construction and Test (1.3). The Construction and Test activity is further subdivided, producing a total of three levels in the hierarchy. In this example, the WBS follows the hierarchical breakdown of the design architecture itself.

Descriptions and deliverables for each activity are provided. The deliverables also serve as checkpoints for monitoring the activity. The identification of checkpoints becomes more important as the complexity and duration of an activity increase. The deliverable items identified for this example include circuit design schematics, simulation results, and test data. Deliverables are even defined for activities that can be somewhat nebulous, such as Research A/D Converters (2.1). Doing so ensures that activities do not become open-ended and have specific deliverables for monitoring their progress.

The fifth column in Table 10.1 is the estimated duration for each activity. The ability to estimate durations is heavily influenced by past experience. The general tendency is to underestimate the amount of time required. Keep in mind that credibility is lost if delivery of the final system is repeatedly delayed, so durations should be estimated as accurately as possible. Take into account time for unexpected problems, such as equipment failure, delayed delivery of parts and equipment, and illness. System integration and testing are tasks that are notoriously time-consuming. A method for estimating activity duration comes from the project evaluation review technique (PERT) developed by the U.S. Navy in the 1950s. Empirical studies show that durations typically follow a beta probability density. From this model, the duration of an activity is estimated as

$$t_e = \frac{t_a + 4t_m + t_b}{6}, \tag{1}$$

where t_a is the most optimistic time estimate, t_b is the most pessimistic time estimate, and t_m is the most realistic time estimate. The advantage of this is that it forces one to consider best- and worst-case scenarios in the plan.

The persons responsible for each deliverable are identified in the WBS. Teams need to consider how to assign responsibility to ensure mutual accountability is achieved. Assigning one person to an activity provides a single person who is responsible for it. However, what if that person needs additional support or backup? Who is going to help if that person is unable to deliver? What if a team member becomes ill? It may make sense to assign multiple people to activities, depending on complexity. In Table 10.1, when a number is placed adjacent to a person's name, it indicates that the person has primary (1) or secondary (2) responsibility for that activity.

The resources needed to complete the activity include material, equipment, and labor. Material includes items that are consumed in creating the deliverable, such as electronic components and printed circuit boards. Examples of equipment include test equipment, computers, and software.

Predecessors for an activity are the activities that must be completed before the given activity can begin. Identifying predecessors is necessary for determining the sequencing of the activities and the time to complete the project.

10.2 Network Diagrams

A *network diagram* is a directed graph representation of the activities and dependencies between them for a project. Activity B is dependent on activity A when A must be completed before B can be started. The network diagram allows a graphic visualization of the project that also allows for quantitative analysis. In the *activity on node* (AON) form, the activities are represented by nodes and the dependencies by arrows. Examples of the basic connections allowed in the AON representation are shown in Figure 10.1.

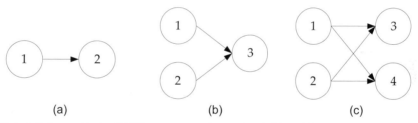

Figure 10.1 Activity on Node (AON) representations of activities. (a) Activity 1 must be completed before activity 2 can begin. (b) Activities 1 and 2 must be completed before activity 3 can begin. (c) Activity 1 and 2 must be completed before activities 3 and 4 can begin.

An example network diagram for a simple project is shown in Figure 10.2. Dummy activity nodes for the start and end of the project have been included. The ID number and estimated duration in days are indicated inside each node. A given network diagram will have multiple paths from the start to the end of the project. A path is any connected sequence of activities from the start node to the end node. In this example, there are four paths to completion: $P_1 = \{1, 4\}$, $P_2 = \{1, 5\}$, $P_3 = \{2, 3, 4\}$, and $P_4 = \{2, 3, 5\}$. The completion of an individual path does not result in completion of the project—all paths (and consequently all activities) must be completed for the entire project to finish.

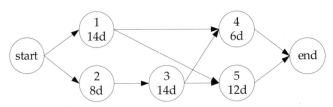

Figure 10.2 Example network diagram. Each activity contains the activity ID and the estimated duration in days.

The duration of each path is determined by summing the duration of all activities on the path, which in this case is 20, 26, 28, and 34 days for paths P_1 to P_4 respectively. The path with the longest duration is known as the *critical path* since it represents the minimum time required to complete the project. In this example the critical path is P_4. Activities on the critical path are of particular interest, since, if they fall behind schedule, or experience *slippage*, the overall completion time of the project is delayed. The other paths can become critical paths if their activities experience sufficient slippage to form a new critical path. A quantity known as *float* quantifies this margin. Float is the amount of time an activity can slip without extending the overall completion time of the project. Thus, all activities on the critical path have zero float by definition. The following method is utilized to determine the float for activities:

1. Identify all paths on the network diagram and the duration for each path.
2. The path with the longest duration is the critical path. Label the critical path duration as t_{cp}. All activities on the critical path have zero float.
3. To determine the float for an activity that is not on the critical path, find all paths that the activity lies on and identify the one with the longest duration. Label this as t_{cp}. The float for the activity is calculated as

$$\text{Float} = t_{cp} - t_{lp}. \qquad (2)$$

Examples 10.1 and 10.2 examine float time computation.

Example 10.1 Float Time Calculation.

Problem: Calculate the float for activities in the network diagram shown in Figure 10.2.

Solution: The first two steps of the process have already been completed, which are identifying the paths and their durations and determining the critical path. To summarize, the paths are $P_1 = \{1, 4\}$, $P_2 = \{1, 5\}$, $P_3 = \{2, 3, 4\}$, and $P_4 = \{2, 3, 5\}$ with durations of 20, 26, 28, and 34 days. The critical path is P_4 and $t_{cp} = 34$ days, so activities 2, 3, and 5 have zero float. Thus

$$\text{Float}_{2,3,5} = \underline{0 \text{ days}}.$$

The only two activities that are not on the critical path are 1 and 4. Let's examine activity 1 first. It lies on paths P_1 and P_2 with durations 20 and 26 days, thus the longest path to completion, t_{lp1}, for activity 1 is 26 days. The float is calculated from (2) as

$$\text{Float}_1 = t_{cp} - t_{lp1} = 34 - 26 \text{ days} = \underline{8 \text{ days}}.$$

Activity 4 lies on P_1 and P_3, so $t_{lp4} = 28$ days and the float is

$$\text{Float}_4 = t_{cp} - t_{lp4} = 34 - 28 \text{ days} = \underline{6 \text{ days}}.$$

This means that activity 1 can slip by 8 days and activity 4 can slip by 6 days without increasing the time to complete the project.

Example 10.2 Network Diagram Construction and Float Time Calculation for the Temperature Display.

Problem: For the example WBS in Table 10.1, (a) create a network diagram, (b) determine the critical path and project completion time, and (c) determine the float for all activities not on the critical path.

Solution:
(a) The network diagram is constructed from the dependencies identified in Table 10.1 and is shown below.

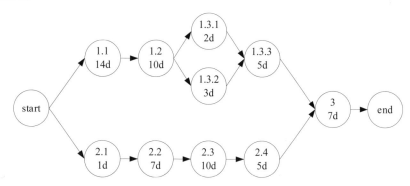

(b) The three paths from start to end are: $P_1 = \{1.1, 1.2, 1.3.1, 1.3.3, 3\}$, $P_2 = \{1.1, 1.2, 1.3.2, 1.3.3, 3\}$, and $P_3 = \{2.1, 2.2, 2.3, 2.4, 3\}$ which have durations of 38, 39, and 30 days respectively. Thus the critical path is P_2, and its duration is

$$t_{cp} = \underline{39 \text{ days}}.$$

(c) All activities on path P_1 are also part of the critical path, with the exception of activity 1.3.2, which has

$$\text{Float}_{1.3.2} = 39 - 38 \text{ days} = \underline{1 \text{ days}}.$$

Activities 2.1–2.4 have the collective float

$$\text{Float}_{2.1-2.4} = 39 - 30 \text{ days} = \underline{9 \text{ days}}.$$

The strength of the network diagram is that it provides an intuitive graphical representation of activities and their dependencies on one another. This is particularly valuable for complex projects where the paths to completion may not be obvious. It also allows identification of the critical path and float times for activities. A disadvantage of the network diagram is that it may be difficult to encapsulate the amount of information required for an in-depth project on a single page in an easy-to-read format.

10.3 Gantt Charts

Gantt charts, developed by a mechanical engineer named Henry Gantt (1861–1919), are a bar graph representation of activities on a timeline. An example Gantt chart is shown in Figure 10.3 for the temperature display design. The Gantt chart effectively shows the WBS and the timeline for completion. A traditional weakness of the Gantt chart has been the inability to show the dependencies between activities. However, as seen in this example, this has been remedied by modern project management software, in which the dependencies are indicated by the connecting arrows between tasks.

Task Name	Start	Finish	Duration
1: Interface Circuitry	1/10/2005	2/22/2005	32d
1.1: Design Circuitry	1/10/2005	1/27/2005	14d
1.2: Purchase Components	1/28/2005	2/10/2005	10d
1.3: Construct & Test Circuits	2/11/2005	2/22/2005	8d
1.3.1: Current Driver Circuitry	2/11/2005	2/14/2005	2d
1.3.2: Level Offset & Gain Circuitry	2/11/2005	2/15/2005	3d
1.3.3: Integrate Components	2/16/2005	2/22/2005	5d
2: LED and Driver Circuitry	1/10/2005	2/9/2005	23d
2.1 Research A/D Converters	1/10/2005	1/10/2005	1d
2.2 Complete Hardware Design	1/11/2005	1/19/2005	7d
2.3 Purchase LED & Driver Components	1/20/2005	2/2/2005	10d
2.4: Construct and Test	2/3/2005	2/9/2005	5d
3: System Integration and Test	2/23/2005	3/3/2005	7d

Figure 10.3 Gantt chart for the temperature display project, created by using Microsoft Visio™.

10.4 Cost Estimation

The second main objective of this chapter is to address how to complete projects within budget. In order to do this, the costs associated with the design, development, and manufacture of the system need to be estimated. This section describes break-even cost analysis and economic considerations followed by methods of cost estimation.

10.4.1 Break-Even Analysis

A break-even analysis aims to determine the number of units that must be sold for costs and revenues to be equal—in other words, for there to be no profit or loss. The two types of costs that factor into this analysis are fixed and variable costs. *Fixed costs* are those that are constant regardless of the number of units produced and cannot be directly charged to a process or activity. Examples are rent, overhead, insurance, property taxes, design and development costs, capital expenditures, market research, and sometimes labor costs, depending upon the situation. Capital expenditures are costs incurred for long-term assets such as equipment or buildings. *Variable costs* vary depending upon the process or items being produced, and fluctuate directly with the number of units produced. Examples are raw materials, inventory, energy costs, and labor costs.

The *break-even point* is the point where the number of units sold is such that there is no profit or loss. It is determined from the total costs and revenue. The total cost required to produce a product is the sum of fixed and variable costs. Assuming n units are sold, the total cost is

$$\text{Total cost} = \text{fixed cost} + n \times \frac{\text{variable cost}}{\text{unit}}. \tag{3}$$

The total revenue generated by the sale of the n units is directly related to the sale price

$$\text{Revenue} = n \times \frac{\text{sale price}}{\text{unit}}. \tag{4}$$

The break-even point is where the revenue and total costs are equal

$$\boxed{n \times \frac{\text{sale price}}{\text{unit}} = \text{fixed cost} + n \times \frac{\text{variable cost}}{\text{unit}}}. \tag{5}$$

The break-even analysis is shown graphically in Figure 10.4 with the costs and revenue plotted as a function of units sold. Equation (5) allows different scenarios to be examined. For example, for a certain cost structure and sales price, the volume of sales necessary to break even can be computed. Or, for a cost structure and projected number of units sold, a target sales price can be selected. Example 10.3 demonstrates the application of break-even analysis for the development of the Hewlett-Packard (HP) DeskJet printer.

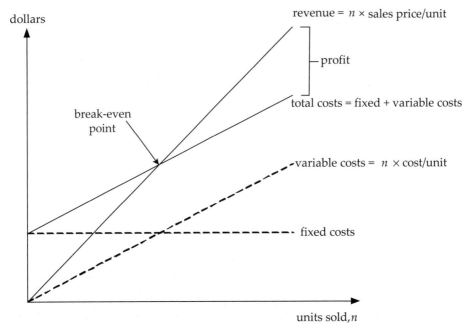

Figure 10.4 Graphical representation of the break-even analysis.

Example 10.3 Break-even Analysis for the HP DeskJet Printer.

Problem: The following data has been publicly reported for the development and sale of the HP DeskJet [Ulr03]: sales price = $300, development cost = $50 million, production investment = $25 million, annual production (sales) volume = 4 million units per year, and sales lifetime = 2 years. Assuming a fictitious variable production cost of $225/unit, determine: (a) the number of units that must be sold to break even and (b) the profit expected over an estimated sales lifetime of 2 years.

Solution:

(a) The objective is to determine the sales volume n necessary to break even. The fixed costs are the sum of the development costs and production investment. So

$$\text{Fixed costs} = \$(50 + 25) \text{ million} = \$75 \text{ million}.$$

This leads to a total cost for n units sold of

$$\text{Total cost} = \$75 \text{ million} + n \times \$225.$$

The revenue is

$$\text{Revenue} = n \times \$300.$$

Setting the revenue and total cost each equal at the break-even point produces

$$n \times \$300 = \$75 \text{ million} + n \times \$225.$$

Solving for the final number of units gives

$$N = \underline{1 \text{ million units}}.$$

(b) Profit is the differential between the total revenue and the total cost and is expressed as

$$\text{Profit} = \text{total revenue} - \text{total costs}.$$

For an expected volume of 4 million units per year over 2 years,

$$\text{Profit} = n \times \$300 - (\$75 \text{ million} + n \times \$225)$$
$$= 8 \text{ million} \times \$300 - (\$75 \text{ million} + 8 \text{ million} \times \$225)$$
$$\text{Profit} = \underline{\$525 \text{ million}}.$$

10.4.2 Cost Models

The costs must be accurately estimated in order to realize the expected profit. The goal here is not to address the complete subject of cost estimation, which is beyond the scope of this book, but instead to present some basic concepts and techniques for estimating development costs. Many projects go over budget during development, and this is a particularly important consideration from a design viewpoint. The WBS is a valuable tool for cost estimation because it divides the project into manageable pieces whose individual costs can be more readily estimated.

As identified in the WBS, the development costs associated with a project typically include labor, equipment, and materials. Equipment costs can be determined in a fairly straightforward manner, because many of the equipment needs are known beforehand. Labor costs are tied directly to the length of the project and are often the largest expense.

Estimates of labor costs are usually based upon past experience and expert opinion. This means asking others to estimate the cost and use it as a guide. The estimation formula in (1) for activity duration can be applied for costs as

$$\text{Cost} = \frac{\text{cost}_a + 4\text{cost}_m + \text{cost}_b}{6}, \tag{6}$$

where cost_a is the most optimistic cost estimate, cost_m is the most likely cost estimate, and cost_b is the most pessimistic cost estimate.

A more formal approach for estimating labor costs is to use empirical models that represent a quantification of past experience. The models estimate an output based upon quantifiable inputs related to the design or technology. Example inputs are the number of subsystems, the estimated complexity of a circuit design, or an estimate of the lines of code necessary for a software project. This changes the problem from opinion-based estimation to estimation of a quantity that is presumably easier to find and a better indicator of the cost. The output of the model is the cost or another quantity that is directly related to it, such as the estimated number of worker-hours.

The simplest example is a linear model ($y = mx + b$) for estimating an output y (cost or person-hours) based upon an input x. For example, IBM modeled software development project costs, using the number of lines of code as the input [Jal97], as

$$\text{Effort} = a \times \text{KLOC} + b. \tag{7}$$

The output is an estimate of the effort in person-months and the input is the projected number of lines of code, KLOC. KLOC, pronounced "kayloc," equals thousands (kilo) of lines of code. The linear model was found to work well for relatively small development projects with between 4 and 10 KLOC. As the complexity increased, an exponential model was found to be more realistic, where

$$\text{Effort} = a(\text{KLOC})^b. \tag{8}$$

By observation of 60 software development projects at IBM, the exponential model was fit to observed data and the parameters were estimated to be $a = 5.2$ and $b = 0.91$. The KLOC model is applied in Example 10.4.

Example 10.4 Effort Estimation Using KLOC.

Problem: Consider a software development project that has a team of 10 software development engineers. The team has proposed a design and estimates that it will require 50,000 lines of code to complete the project. The average cost to the company for an engineer is $100,000 per year, including salary, benefits, and overhead. Estimate (a) the time required to the complete the project and (b) the labor costs.

Solution:
(a) For the projected value KLOC, the exponential model in (8) is most appropriate:

$$\text{Effort} = 5.2(50)^{0.91} = 183 \text{ worker-months.}$$

Since there are 10 developers on the project, the estimated time is determined by dividing the effort by 10, to produce an estimated time of 18.3 months.

(b) The labor costs for development are determined from the number of worker-months and the average monthly salary of a development engineer as

$$\text{Labor cost} = 183 \text{ worker-months} \times \frac{\$100,000}{\text{year}} \times \frac{1 \text{ year}}{12 \text{ months}}$$

$$\text{Labor cost} = \$1.53 \text{ million.}$$

The models in (7) and (8) are simplistic in that there is a single input to the estimator. For example, what if there were 100 engineers assigned to the project in Example 10.4? The model indicates that the project would be completed in 1.83 months at the same cost. Common sense dictates that this is unrealistic; this problem is commonly referred to as the "mythical man-month" (worker-month). The mythical worker-month refers to the fact that just adding more people to the team will not linearly reduce development time. The reality is that many factors affect the costs, which leads to effort estimation based on many inputs. An example of this is the constructive cost model (COCOMO) that is also used in software development. There are different levels of COCOMO that allow increasingly complex model inputs, such as the type of technology employed, the maturity of the technology, the size of the team, the experience of the engineers, and the time frame required to complete the project. Not only do the models estimate costs, but they can also be used to estimate how many engineers are needed to complete projects of a given complexity. Although the examples cited here are from the software field, the concepts are general and applicable to the development of many technologies.

Empirical cost models can also be developed for the materials necessary for the manufacture of items. Material estimates are often based upon the expected size and types of technologies used in the manufactured part. For example, a cost model for the manufacture of a printed circuit board may have as inputs the size of the board, the number of layers, and the type of technology used (such as through-hole versus surface mount).

10.5 The Project Manager

Many engineering teams have a project manager responsible for planning and organizing the project. The project manager may take primary responsibility for developing the WBS, the network diagram and Gantt chart, the cost estimates, and the budget. Although the project manager may have primary responsibility for the project plan, all team members should have input and contribute to development of the plan. The project manager should monitor the checkpoints and deliverables against the plan and develop strategies for reacting to slippage in any activities. The plan should be updated as necessary and the changes communicated to all involved. The project manager may also have primary responsibility for the purchasing of materials and controlling spending. Project managers are not necessarily the boss in the traditional sense and should be viewed as a member of the team. As such, it is important for the project manager to also be responsible for completing project deliverables in addition to the project management tasks.

10.6 Guidance

The following is guidance to be considered when you are creating the project management plan, and is particularly relevant for those working on capstone design projects:

- *Build the plan after the design architecture is complete.* The project plan can be created at any point in the design process. Our experience shows that a good time to develop it is after the system design architecture is complete. The design serves as a good guide for developing the WBS.

- *Take the initial time estimates for activities and double them!* Most people tend to significantly underestimate the amount of time it takes to complete an activity. That is because people often have a conceptual idea of what it will take to complete the task and can envision the steps to completion. The desire to please superiors also influences people to underestimate the time to completion. Although it may not be necessary to double the time estimates, you can incorporate the most optimistic and pessimistic estimates and apply equation (1). Or, if a proven mathematical model exists, such as the KLOC estimator in equations (7) and (8), it can be used to estimate times.
- *Assign a lot of time for testing and integration.* During integration many people must work together to integrate components that may have been developed in isolation. Problems with a single component can bring the integration to a halt. Delays may be compounded by necessary redesign to correct the problem.
- *Factor in lead times for part ordering.* Even with the Internet and overnight delivery, you may find that needed parts and equipment are out of stock. Lead times for seemingly commonplace items can sometimes be quite lengthy.
- *Assign a project manager.* Consider assigning one individual who has primary responsibility for organizing and monitoring the plan. Again, the project manager must also be responsible for some of the deliverables for completion of the project.
- *Do not assign all team members to all tasks.* Experience shows that when this is the case nobody is responsible for anything and the work doesn't get done. There needs to be individual accountability for all team members. However, it may be a good idea to have more than one person responsible for activities for backup support as shown in the WBS in Table 10.1.
- *Track the progress versus the plan.* There is a tendency to create the plan and then ignore it. The plan is valuable only if it is monitored and progress is tracked.
- *Don't become a slave to the plan.* Circumstances usually dictate change. Be prepared to shift resources as needed. Monitor the plan to see if there are changes to the critical path or if a new critical path emerges.
- *Experience counts.* Get started now in developing this experience by creating a plan for your project.

10.7 Project Application: The Project Plan

A project plan should contain the following:

- *Work Breakdown Structure.* Identify the activities, deliverables, responsibilities, duration, resources, and dependencies as demonstrated in Table 10.1. Be sure to provide sufficient detail in the structure and identify clear deliverables.
- *Gantt Chart and/or Network Diagram.* Provide a graphical representation of the project plan. Network diagrams have the advantage of showing the dependencies, while Gantt charts show the time frame. Modern software tools allow both to be integrated into the same graph and are a good compromise as demonstrated in Figure 10.3. The critical path should be identified and the float for non-critical-path activities understood.
- *Costs.* Develop a tabulated list of costs and for the equipment, materials, and labor necessary to carry out the project. It may not be necessary to develop labor costs for a capstone project, yet it is good practice to estimate worker-hours and compare the estimate to the actual at the end of the project.

10.8 Summary and Further Reading

Three main objectives of project management are to complete projects that are on time, within budget, and meet the needs of the user. This chapter addressed the time and budget aspects. The key element in developing a project plan is the WBS, which is a hierarchical identification of the activities needed to complete the project. Both network diagrams and Gantt charts can be created from the WBS. A network diagram is a graphical representation of activities and their dependencies that provides for quantitative analysis of the project plan. This analysis includes computation of the critical path and float times. The Gantt chart is related to the network diagram, but provides a time scale representation of the activities. In terms of the budget issues, a simple profit-and-loss model was presented with a break-even analysis. Model-based techniques can be used for estimating labor costs, where the models are built from the analysis of similar projects. The models can be linear or nonlinear with single inputs. More complex models can be developed with multiple inputs in order to arrive at a more precise estimate.

Project management is a well-developed field and there are many good textbooks and online resources available for delving deeper into the subject. Project Management by Gray and Larson [Gra02] is a comprehensive text on the subject that includes risk management, resource scheduling, leadership, and performance measurement. Planning, Performing, and Controlling Projects by Angus et al. [Ang00] is written for an engineering and scientific audience, and integrates phases of the design process. MindTools (www.mindtools.com) is an online resource that addresses many career skills including project management.

10.9 Problems

10.1 In your own words, describe what is meant by the work breakdown structure.

10.2 Consider the set of activities, duration (in days), and predecessors for a project given below.

Activity	A	B	C	D	E	F	G	H	I
Duration	3	9	6	6	6	3	2	6	7
Predecessors	-	-	-	A, B	D, B	C	F, E	G	F

a) Develop a network diagram representation for the project.
b) Determine the critical path.
c) Determine the float time for all activities that are not on the critical path.

10.3 Consider the set of activities, duration (in days), and predecessors for a project given below.

Activity	A	B	C	D	E	F	G	H	I	J	K
Duration	9	12	3	4	5	9	8	3	6	9	1
Predecessors	-	A	A	B, C	C	B	D	F, D	G	H, I	E

a) Develop a network diagram representation for the project.
b) Determine the critical path.
c) Determine the float time for all activities that are not on the critical path.

10.4 Explain why a network diagram cannot contain cycles. A cycle is a sequence of activities where you can travel back to an activity already visited.

10.5 Describe the advantages and disadvantages of the network diagram and Gantt chart representations for a project.

10.6 Assume that the following data has been determined for the development and sale of a new digital thermometer for home use: development cost = $250,000, production investment = $500,000, annual production volume = 20,000 units per year, and sales lifetime = 7 years. Assuming a variable production cost of $5 per unit, determine: (a) the sales price necessary to break even within 2 years, and (b) the profit expected over the estimated sales lifetime.

10.7 Describe the difference between the cost estimation models in equations (7) and (8) and the COCOMO cost estimation model.

10.8 Consider a software development project that has a team of 50 software development engineers. The team has proposed a design and estimates that it will require

500,000 lines of code to complete the project. The average cost to the company for an engineer is $90,000 per year, including salary, benefits, and overhead. Estimate (a) the time required to complete the project, and (b) the labor costs.

10.9 Consider a software development project where the team has proposed a design and estimates that it will require 200,000 lines of code to complete. The average cost to the company for an engineer is $110,000 per year, including salary, benefits, and overhead. Estimate (a) the number of engineers needed to complete the project within 18 months, and (b) the labor costs.

10.10 **Project Application.** Develop a project plan for your project. A format and guideline for developing the plan is contained in Section 10.7.

Chapter 11 Ethical and Legal Issues

A man without ethics is a wild beast loosed upon this world.—Albert Camus

In your professional career you are going to face ethical dilemmas, and how you respond to them will affect your reputation, your employability, and the welfare of others. Systems are developed for use by other people, and the decisions that go into their design affect the health and safety of those people. There are plenty of high-profile examples, such as the two space shuttle disasters, that make it clear that the decisions of engineers and scientists have serious implications. Beyond issues in the technical aspects of design, there are professional ethics that govern the broad scope of interactions between people in the workplace. The aim of this chapter is to present the basics of engineering ethics and provide guidance for addressing dilemmas when they arise. It presents a basic overview of morality and ethics, professional codes that apply to the engineering profession, intellectual property and legal issues as they relate to design, how to apply ethics throughout the design process, and guidance for handling ethical dilemmas. The chapter concludes with case studies that provide an opportunity to apply ethical decision-making skills.

Learning Objectives

By the end of this chapter, the reader should:
- Understand what is meant by morals, ethics, and values.
- Be familiar with the IEEE Code of Ethics.
- Understand what a patent is, the criteria for filing for one, and the elements that constitute it.
- Understand the difference between patents, trade secrets, and copyrights.
- Understand the concepts of negligence and strict liability as they apply to product design.
- Understand how to incorporate ethical issues throughout the design process.
- Be able to analyze ethical case studies and suggest solutions to the dilemmas that they embody.

11.1 Ethical Theory in a Nutshell

Here is an ethical dilemma to consider as we start our discussion. Let's assume that you are conducting a job search during your senior year in college and have interviewed with several prospective employers. Company A offers you a job and you accept the position and sign a contract agreeing upon a starting salary, position, and start date. Two weeks later, Company B also offers you a job with a higher starting salary and work that you find personally more interesting. You have a choice to make. Do you turn down the offer of Company B and stay with A? Or do you accept the offer made by B and then inform A that you are not going to work for them? This is a classic dilemma that many students will face, and rather than try to answer it now, let's revisit it later.

There is often confusion as to what exactly is meant by the related concepts of ***ethics*** and ***morals***. Morality is concerned with principles of right and wrong and the decisions that derive from these principles. Morals are often taught by stories that are common in all cultures, such as Aesop's fable of the boy who cried wolf. In this story, the boy was watching sheep and cried wolf when there was no wolf present. In response, all of the villagers came running to help him, only to find no wolf. When the wolf did later appear and the boy cried for help once again, nobody believed him nor came to help, and the wolf scattered the flock. The moral of this story is that it is wrong to lie, and when people become known as liars, nobody will believe them even if they are telling the truth.

The term *ethics* is closely related to morals and they are often used interchangeably. According to the American Heritage Dictionary, ethics is defined as:

> 1. Branch of philosophy that deals with the general nature of good and bad and the specific moral obligations of and choices to be made by the individual in her/his relationship to others. 2. Rules or standards governing conduct, especially those of a profession.

Ethics is the philosophy or study of moral obligations and the choices to be made by individuals. It is important to understand that if there is no decision to be made, there is no ethical dilemma. The choices to be made are based upon a belief as to what is good or bad, and the decisions must impact relationships to others.

Morals derive from ***principles,*** which are fundamental laws or rules that govern behavior. An example of a principle is the Golden Rule, which states that people should treat others as they themselves would like to be treated. The Golden Rule is a universal principle and embodied in one form or another in all major religions and belief systems of the world. Another example of a principle is the belief that people should be honest and trustworthy in their dealings with others. Value is another term that is often heard in ethical discussions. A ***value*** is something that a person or group believes to be valuable or worthwhile. This could be relatively innocuous such as valuing baseball as a sport, or something more significant such as valuing hard work. A group of thieves may believe that it is valuable to steal from other people, but not from their own group. As these examples show, shared values may be good or bad. The final example clearly violates the principles of the Golden Rule and is considered bad by most people.

Rule-based ethics are based upon a set of rules that can be applied to make decisions. In the strictest form they are considered to be absolute in terms of governing behavior—either you follow the rule (good) or break the rule (bad). This type of an ethical system is based upon the principles of *universality* and *transitivity*. Universality means that the governing rules are such that they can be accepted by everyone, while transitivity means that you would accept others applying the same decision to you. A problem with rule-based ethics is defining a universal set of rules that everyone can agree upon. There may be a few rules or principles that can be agreed upon by all, but going beyond that is difficult.

Conditional rule-based ethics means that there are certain conditions under which an individual can break a rule. This is generally because it is believed that the moral good of the situation outweighs the rule. For example, if you have a seriously injured person in your car and are transporting them to the hospital, is it acceptable to exceed the speed limit? In this case, it may be deemed that the moral good of getting the person prompt medical attention outweighs the obligation to obey the speed limit. It is generally believed that killing others is wrong, except in the case of war. Is it acceptable to cheat on an exam because you stayed up all night to help your sick roommate? Or is it acceptable to cheat because you simply did not have enough time to study because of your other obligations? In each of these cases a moral choice has to be weighed.

In **utilitarian ethics**, decisions are based upon the decision that brings about the highest good for all, relative to all other decisions. This sounds appealing, but has drawbacks in that bad choices may need to be made for certain parties in order to achieve this overall good. It becomes very difficult to determine exactly what the highest good is. *Situational ethics* bases decisions on whether they produce the highest good for the person—that is, decisions based upon their effect on the individual and situation at hand. This is generally considered a poor ethical decision-making approach.

Let's conclude this section with a new scenario. Let's assume that you are conducting a job search and Company A offers you a job, you agree upon terms of the offer, accept it, and sign a contract. You then inform the other companies that interviewed you that you are no longer available. One month later Company A calls you and notifies you that they are rescinding their offer. Although they can no longer offer you the job, they will provide you with $1,000 in compensation for rescinding the offer. You later find out that Company A just offered the exact same job to another student in your class who has a higher grade point average than yours and more experience with the technical products the company designs. Was the decision of the company ethical? Returning to the original case, is it ethical for the student to rescind his or her acceptance of the original job offer and then accept the other?

11.2 The IEEE Code of Ethics

Most engineering disciplines have associated with them a professional society or organization. In electrical and computer engineering two important ones are the Institute of Electrical and Electronics Engineers (IEEE) and the Association for Computing Machinery (ACM). The objective of professional societies is to promote and support their respective fields. They offer a variety of services and benefits to their members, such as access to technical information, networking opportunities, and financial services. They also define accepted practices of their members that are embodied in an ethical code. In fact, professional societies were originally created to provide guidance for ethical practices in their field and to ensure the safety of the public. The IEEE Code of Ethics, shown in Table 11.1, applies broadly to the electrical, computing, and software fields.

Table 11.1 The IEEE Code of Ethics.

We, the members of the IEEE, in recognition of the importance of our technologies in affecting the quality of life throughout the world, and in accepting a personal obligation to our profession, its members and the communities we serve, do hereby commit ourselves to the highest ethical and professional conduct and agree:

1. to accept responsibility in making engineering decisions consistent with the safety, health, and welfare of the public, and to disclose promptly factors that might endanger the public or the environment;

2. to avoid real or perceived conflicts of interest whenever possible, and to disclose them to affected parties when they do exist;

3. to be honest and realistic in stating claims or estimates based on available data;

4. to reject bribery in all its forms;

5. to improve the understanding of technology, its appropriate application, and potential consequences;

6. to maintain and improve our technical competence and to undertake technological tasks for others only if qualified by training or experience, or after full disclosure of pertinent limitations;

7. to seek, accept, and offer honest criticism of technical work, to acknowledge and correct errors, and to credit properly the contributions of others;

8. to treat fairly all persons regardless of such factors as race, religion, gender, disability, age, or national origin;

9. to avoid injuring others, their property, reputation, or employment by false or malicious action;

10. to assist colleagues and co-workers in their professional development and to support them in following this code of ethics.

Approved by the IEEE Board of Directors, August 1990

Some of the values embodied in the code are that of treating others with fairness and respect (8–10), honesty and trustworthiness (2–4), and professional competence (6–7). The importance of the safety of the public is clearly identified in the first and ninth items of the code. The code provides a common basis for analyzing ethical case studies and is applied for these purposes in subsequent sections.

11.3 Intellectual Property and Legal Issues

It is important to understand some legal issues, particularly as they impact product design and development. The first considered is intellectual property, which seeks to answer the question of who owns an idea or invention. The second is that of legal liability, which comes into play if somebody is injured or harmed by a product or system.

11.3.1 Intellectual Property

The intent of designing a new product is usually to sell it for a profit, which leads to the issue of ownership of both the intellectual property and the profits. The tools for protecting intellectual property are patents, trade secrets, and copyrights. Be aware that when you enter the workforce as an engineer, your employer may ask you to sign an agreement assigning the company rights to all of the intellectual property that you create while in their employ. Such contracts are common and enforceable.

The most well-known way of protecting a design or invention is with a *patent*. If a patent is held for a technology, others cannot use it without permission of the owner. The owner can deny others the right to use it, or grant the right to use it in exchange for monetary compensation. The two types of patents are utility and design patents. The United States Patent and Trademark Office (USPTO) defines the difference between the two as follows:

> *A utility patent may be granted to anyone who invents or discovers any new and useful process, machine, article of manufacture, compositions of matter, or any new useful improvement thereof. A design patent may be granted to anyone who invents a new, original, and ornamental design for an article of manufacture.*

The one that is of most interest for this discussion is the utility patent, as the design patent focuses on aesthetic design issues.

In order to be granted a utility patent, the invention must meet three conditions: it must be novel, nonobvious, and useful. Novelty means that the idea must be new and that nothing like it already exists. It cannot be an idea that is already patented or has been published for more than a year. The nonobvious condition means that another person would not be expected to develop the same idea based upon existing technology. As an example, for personal computers, it would not be possible to patent the idea of increasing data storage by adding multiple disk drives, since this is an obvious extension of existing technology. Useful means that the device must perform a useful function and be able to be reduced to practice.

You could not patent the concept for a transporter of the type seen on science fiction shows, such as *Star Trek*, where people are beamed instantaneously through space from one place to another—unless of course you can reduce it to practice.

To a file a patent, extensive research needs to be done to be sure that the idea is novel. A good place to start is the U.S. Patent Trademark Office (USPTO) website (www.uspto.gov) and its searchable database of patents back to 1790. The database allows full text search of patents from 1976 forward and patent number search back to 1790. It provides full images of all patents in the database. The website also has information on how to apply for a patent and describes the differences between patents, trademarks, and copyrights. The elements that are contained in a patent are:

- *A citation of prior art.* This lists similar patents and publicly reported technology.
- *A description of the invention.* This describes how it operates and how it would be reduced to practice.
- *Claims.* They are the legal description of the invention and its unique aspects.

The claims are used to determine if another party is infringing upon a patent, and thus they must be carefully thought out and properly worded. If the claims are too broad or too specific then they may not provide much protection. There are fees for filing the patent and for periodic maintenance that can cost between $5,000 and $10,000 over the life of the patent. Hiring a good patent attorney is invaluable in the application process and will cost money—it is not unusual to spend $20,000 for lawyer and application fees. When selecting a patent attorney, make sure that you choose one who has a good track record and reputation.

An example patent for a hardware design of a fast floating point overflow and sign detection unit (U.S. Patent 6,779,013) is shown in Figure 11.1(a). The patent is owned by Intel Corporation and the inventor was one of their employees. The first page identifies prior art that relates to this patent. This includes previous patents related to this technology, which in this case goes back to 1991. It is not unusual to cite prior patents that go back 50 or 100 years. In addition to patents, prior art in publicly available literature is identified. The first page also identifies the number of claims and supporting figures, and shows the high-level design architecture. A description of the system starts on the second page as shown in Figure 11.1(b). The complete description is fairly lengthy and only the first page is shown. It is similar to a technical report in that it provides background on the technology and describes the invention and its operation in detail. The final page of the patent (Figure 11.1(c)) identifies the legal claims that make the invention unique.

Chapter 11 Ethical and Legal Issues

(12) **United States Patent**
Pangal

(10) Patent No.: **US 6,779,013 B2**
(45) Date of Patent: **Aug. 17, 2004**

(54) **FLOATING POINT OVERFLOW AND SIGN DETECTION**

(75) Inventor: **Amaresh Pangal**, Hillsboro, OR (US)

(73) Assignee: **Intel Corporation**, Santa Clara, CA (US)

(*) Notice: Subject to any disclaimer, the term of this patent is extended or adjusted under 35 U.S.C. 154(b) by 368 days.

(21) Appl. No.: **09/873,744**

(22) Filed: **Jun. 4, 2001**

(65) **Prior Publication Data**
US 2002/0194239 A1 Dec. 19, 2002

(51) Int. Cl.[7] G06F 7/44
(52) U.S. Cl. **708/503**
(58) Field of Search 708/503, 501

(56) **References Cited**

U.S. PATENT DOCUMENTS

5,764,089 A		6/1998	Partovi et al. 327/200
5,898,330 A		4/1999	Klass 327/210
5,900,759 A		5/1999	Tam 327/201
5,993,051 A	*	11/1999	Jiang et al. 708/501
6,205,462 B1	*	3/2001	Wyland et al. 708/503
6,360,189 B1	*	3/2002	Hinds et al. 708/501
6,480,872 B1	*	11/2002	Choquette 708/501

OTHER PUBLICATIONS

Beaumont–Smith, A., et al., "Reduced Latency IEEE Floating–Point Standard Adder Architectures", *Proceedings of the 14th IEEE Symposium on Computer Arithmetic*, 8 pgs., (1998).

Even, G., et al., "On the Design of IEEE Compliant Floating Point Units", *IEEE Transactions on Computers*, vol. 49, 398–413, (May 2000).

Goto, G., et al., "A 54×54–b Regularly Structured Tree Multiplier", *IEEE Journal of Solid–State Circuits*, vol. 27, 1229–1236, (Sep. 1992).

Ide, N., et al., "2.44–GFLOPS 300–MHz Floating–Point Vector–Processing Unit for High–Performance 3–D Graphics Computing", *IEEE Journal of Solid–State Circuits*, vol. 35, 1025–1033, (Jul. 2000).

Klass, F., "Semi–Dynamic and Dynamic Flip–Flops with Embedded Logic", *Proceedings of the Symposium on VLSI Circuits, Digest of Technical Papers*, Honolulu, HI, IEEE Circuits Soc. Japan Soc. Appl. Phys. Inst. Electron., Inf. & Commun. Eng. Japan, pp. 108–109, (1998).

Lee, K.T., et al., "1 GHz Leading Zero Anticipator Using Independent Sign–Bit Determination Logic", *2000 Symposium on VLSI Circuits Digest of Technical Papers*, 194–195, (2000).

Partovi, H., et al., "Flow–Through Latch and Edge–Triggered Flip–Flop Hybrid Elements", *Proceedings of the IEEE International Solid–State Circuits Conference, Digest of Technical Papers and Slide Supplement*, NexGen Inc., Milpitas, CA, 40 pgs., (1996).

(List continued on next page.)

Primary Examiner—Tan V. Mai
(74) *Attorney, Agent, or Firm*—Schwegman, Lundberg, Woessner & Kluth, P.A.

(57) **ABSTRACT**

A multiply-accumulate circuit includes a compressor tree to generate a product with a binary exponent and a mantissa in carry-save format. The product is converted into a number having a three bit exponent and a fifty-seven bit mantissa in carry-save format for accumulation. An adder circuit accumulates the converted products in carry-save format. Because the products being summed are in carry-save format, post-normalization is avoided within the adder feedback loop. The adder operates on floating point number representations having exponents with a least significant bit weight of thirty-two, and exponent comparisons within the adder exponent path are limited in size. Variable shifters are avoided in the adder mantissa path. A single mantissa shift of thirty-two bits is provided by a conditional shifter.

20 Claims, 10 Drawing Sheets

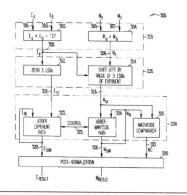

Figure 11.1 (a) Example of a U.S. patent for a floating point overflow and sign detection unit.

US 6,779,013 B2

FLOATING POINT OVERFLOW AND SIGN DETECTION

FIELD

The present invention relates generally to floating point operations, and more specifically to floating point multiply accumulators.

BACKGROUND

Fast floating point mathematical operations have become an important feature in modern electronics. Floating point units are useful in applications such as three-dimensional graphics computations and digital signal processing (DSP). Examples of three-dimensional graphics computation include geometry transformations and perspective transformations. These transformations are performed when the motion of objects is determined by calculating physical equations in response to interactive events instead of replaying prerecorded data.

Many DSP operations, such as finite impulse response (FIR) filters, compute $\Sigma(a_i b_i)$, where i=0 to n−1, and a_i and b_i are both single precision floating point numbers. This type of computation typically employs floating point multiply accumulate (FMAC) units which perform many multiplication operations and add the resulting products to give the final result. In these types of applications, fast FMAC units typically execute multiplies and additions in parallel without pipeline bubbles. One example FMAC unit is described in: Nobuhiro et al., "2.44-GFLOPS 300-MHz Floating-Point Vector Processing Unit for High-Performance 3-D Graphics Computing," IEEE Journal of Solid State Circuits, Vol. 35, No. 7, July 2000.

The Institute of Electrical and Electronic Engineers (IEEE) has published an industry standard for floating point operations in the ANSI/IEEE Std 754–1985, *IEEE Standard for Binary Floating-Point Arithmetic*, IEEE, New York, 1985, hereinafter referred to as the "IEEE standard." A typical implementation for a floating point FMAC compliant with the IEEE standard is shown in FIG. 1. FMAC **100** implements a single precision floating point multiply and accumulate instruction "D=(A×B)+C," as an indivisible operation. As can be seen from FIG. 1, fast floating point multipliers and fast floating point adders are both important ingredients to make a fast FMAC.

Multiplicands A and B are received by multiplier **110**, and the product is normalized in post-normalization block **120**. Multiplicands A and B are typically in an IEEE standard floating point format, and post-normalization block **120** typically operates on (normalizes) the output of multiplier **110** to make the product conform to the same format. For example, when multiplicands A and B are IEEE standard single precision floating point numbers, post-normalization block **120** operates on the output from multiplier **110** so that adder **130** receives the product as an IEEE standard single precision floating point number.

Adder **130** adds the normalized product from post-normalization block **120** with the output from multiplexer **140**. Multiplexer **140** can choose between the number C and the previous sum on node **152**. When the previous sum is used, FMAC **100** is performing a multiply-accumulate function. The output of adder **130** is normalized in post-normalization block **150** so that the sum on node **152** is in the standard format discussed above.

Adder **130** and post-normalization block **150** can be "non-pipelined," which means that an accumulation can be performed in a single clock cycle. When non-pipelined, adder **130** and post-normalization block typically include sufficient logic to limit the frequency at which FMAC **100** can operate, in part because floating point adders typically include circuits for alignment, mantissa addition, rounding, and other complex operations. To increase the frequency of operation, adder **130** and post-normalization block **150** can be "pipelined," which means that registers can be included in the data path to store intermediate results. One disadvantage of pipelining is the introduction of pipeline stalls or bubbles, which decrease the effective data rate through FMAC **100**.

For the reasons stated above, and for other reasons stated below which will become apparent to those skilled in the art upon reading and understanding the present specification, there is a need in the art for fast floating point multiply and accumulate circuits.

BRIEF DESCRIPTION OF THE DRAWINGS

FIG. **1** shows a prior art floating point multiply-accumulate circuit;

FIG. **2** shows an integrated circuit with a floating point multiply-accumulate circuit;

FIG. **3** shows the exponent and mantissa paths of a floating point multiply-accumulate circuit;

FIG. **4** shows a mantissa multiplier circuit;

FIG. **5** shows a floating point conversion unit;

FIG. **6** shows a carry-save negation circuit;

FIG. **7** shows a base 32 floating point number representation;

FIG. **8** shows an exponent path of a floating point adder;

FIG. **9** shows a mantissa path of a floating point adder;

FIG. **10** shows an overflow detection circuit;

FIG. **11** shows a post-normalization circuit; and

FIG. **12** shows a sign detection circuit.

DESCRIPTION OF EMBODIMENTS

In the following detailed description of the embodiments, reference is made to the accompanying drawings which show, by way of illustration, specific embodiments in which the invention may be practiced. In the drawings, like numerals describe substantially similar components throughout the several views. These embodiments are described in sufficient detail to enable those skilled in the art to practice the invention. Other embodiments may be utilized and structural, logical, and electrical changes may be made without departing from the scope of the present invention. Moreover, it is to be understood that the various embodiments of the invention, although different, are not necessarily mutually exclusive. For example, a particular feature, structure, or characteristic described in one embodiment may be included within other embodiments. The following detailed description is, therefore, not to be taken in a limiting sense, and the scope of the present invention is defined only by the appended claims, along with the full scope of equivalents to which such claims are entitled.

Floating Point Multiply Accumulator

FIG. **2** shows an integrated circuit with a floating point multiply-accumulate circuit. Integrated circuit **200** includes floating point multiplier **210**, floating point conversion unit **220**, floating point adder **230**, and post-normalization circuit **250**. Each of the elements shown in FIG. **2** is explained in further detail with reference to figures that follow. In this section, a brief overview of the FIG. **2** elements and their

Figure 11.1(b) Second page of the patent description.

US 6,779,013 B2

S1	C1	MC	Sign
1	0	0	0
1	0	1	1
1	1	X	-

Magnitude comparator **325** operates in parallel with adder mantissa path **324**, so MC is available for sign detection circuit **1104** at substantially the same time as Msum. In this manner, the operation of sign detection circuit **1104** does not appreciably increase the delay within the feedback loop.

CONCLUSION

The method and apparatus of the present invention provide a fast multiply-accumulate operation that can be made compliant with any floating point format. Furthermore, the method and apparatus of the present invention can provide precision comparable to the precision available using prior art double precision arithmetic units, in part because the mantissa fields are expanded. In some embodiments, IEEE standard single precision operands are multiplied and the products are summed. The multiplier includes a compressor tree to generate a product with a binary exponent and a mantissa in carry-save format. The product is converted into a number having a three bit exponent and a fifty-seven bit mantissa in carry-save format for accumulation. An adder circuit accumulates the converted products in carry-save format. Because the products being summed are in carry-save format, post-normalization is avoided within the adder feedback loop. In addition, because the adder operates on floating point number representations having exponents with a least significant bit weight of thirty-two, exponent comparisons within the adder exponent path are fast, and variable shifters can be avoided in the adder mantissa path. When the adder is not pipelined, a fast single cycle accumulation is realized with the method and apparatus of the present invention.

It is to be understood that the above description is intended to be illustrative, and not restrictive. Many other embodiments will be apparent to those of skill in the art upon reading and understanding the above description. The scope of the invention should, therefore, be determined with reference to the appended claims, along with the full scope of equivalents to which such claims are entitled.

What is claimed is:

1. A floating point multiply-accumulate circuit comprising:
   an exponent path including:
      an exponent summer to sum two input exponents having a first weight to produce a product exponent;
      an exponent conversion unit coupled to the output of the exponent summer, to convert the product exponent to a second weight; and
      an exponent accumulation stage to accumulate the converted product exponent and to choose a larger exponent from the converted product exponent and an accumulated exponent; and
   a mantissa path including:
      a mantissa multiplier to multiply two input mantissas and produce a product mantissa in carry-save format that includes a sum field and a carry field;
      a mantissa shifter to shift the sum field and the carry field responsive to the exponent conversion unit in the exponent path; and
      a mantissa accumulator to accumulate shifted product mantissas in carry-save format, the mantissa accumulator including an overflow detection circuit responsive to two most significant bits of a sum field output from the mantissa accumulator.

2. The floating point multiply-accumulate circuit of claim 1 wherein the overflow detection circuit comprises an exclusive-or gate.

3. The floating point multiply-accumulate circuit of claim 1 wherein the exponent conversion unit is configured to zero the least significant five bits of the product exponent.

4. The floating point multiply-accumulate circuit of claim 1 wherein the mantissa shifter is configured to shift the sum field and carry field of the product mantissa by a number of bit positions equal to a value of the least significant five bits of the product exponent.

5. The floating point multiply-accumulate circuit of claim 1, wherein the mantissa accumulator comprises four-to-two compressors.

6. The floating point multiply-accumulate circuit of claim 1 further comprising a post-normalization stage to produce a normalized floating point resultant.

7. The floating point multiply-accumulate circuit of claim 6 wherein the post-normalization stage includes a sign detection circuit.

8. The floating point multiply-accumulate circuit of claim 7 further comprising a magnitude comparator in parallel with the mantissa accumulator, wherein the sign detection circuit is responsive to the magnitude comparator.

9. The floating point multiply-accumulate circuit of claim 6 wherein the post-normalization stage is configured to be turned off until accumulation is complete.

10. The floating point multiply-accumulate circuit of claim 1 wherein the exponent conversion unit is configured to convert the product exponent to have a least significant bit weight equal to thirty-two.

11. A method performed within a programmable digital computer, comprising:
   multiplying two floating point mantissas and summing two floating point exponents to form a product;
   converting the product to have a different least significant bit weight exponent field;
   accumulating the converted product in carry-save format;
   detecting overflow as a function of two most significant bits of a sum field of an accumulated product; and
   post-normalizing the accumulated product.

12. The method of claim **11** wherein accumulating the product comprises adding a first plurality of products with a last product, the method further comprising turning off post-normalization until the last product is accumulated.

13. The method of claim **11** wherein converting comprises:
   shifting a mantissa of the product by an amount equal to the value of the least significant five bits of the exponent of the product; and
   zeroing the least significant five bits of an exponent of the product.

14. The method of claim **11** wherein accumulating comprises:
   comparing an exponent of a first converted product to an exponent of a second converted product;
   conditionally shifting right by a fixed amount the mantissa of the converted product having a smaller exponent;
   selecting the larger exponent as a resultant exponent; and
   producing a resultant mantissa from a mantissa of the first converted product and a mantissa of the second converted product.

Figure 11.1(c) Conclusion of the patent and its claims.

In the United States, patent grants are based upon the concept of first to conceive the invention, not first to file the patent application. If two parties are applying for a patent at the same time, then the one who demonstrates that he or she was the first to conceive the idea and reduce it to practice receives the patent. Good records must be maintained to prove this. It is done by recording inventions in a bound design notebook with numbered pages that cannot be removed. All entries should be clearly described, understandable by other engineers, and be signed and dated. For the best protection, entries should be signed and witnessed by at least one other person and the notebook should be occasionally notarized. Pages should not be removed from the notebook and entries should be made in nonerasable pen. Mistakes and blank spaces should be crossed out, signed, and dated. Figures should be drawn directly in the notebook, while computer-generated figures are pasted in so that they are not removable. It is good practice to maintain a design notebook even if you are not planning to patent your ideas.

Once a patent is granted, it is good for 20 years—after it expires, the invention is fair game for anybody to use. Let's assume that you are the owner of a patent and you find that somebody is infringing upon it. Is the government going to come in and start legal action against the infringer? No—a patent only gives the owner a right to sue others if they infringe upon it. The government does not actively seek out offenders and protect your intellectual property. The owner of the patent must protect it.

There are drawbacks to patents. The owner must be vigilant in defending a patent and may have to initiate legal action to do so. Once an idea is patented, it is made public for all to see on the government's website. This may not be a good idea, depending upon the technology and competitive situation. An alternative that many companies employ to protect intellectual property is to hold them as *trade secrets*. Obviously, the idea must be kept secret so that others cannot find out about it. This is often done by restricting the number of people who have access to the idea and by having those who do know about it sign a *nondisclosure agreement.* It is common practice for companies to ask employees and visitors to their facilities to sign a nondisclosure agreement that prevents the signer from disseminating information about their products, services, and trade secrets. One who breaks a nondisclosure agreement can be held legally liable. Trade secrets pose another type of risk to a company, since once the secret is revealed, it is fair game for competitors. One way to determine a competitor's trade secrets is through the use of *reverse engineering,* where a device or process is taken apart to understand how it works. Reverse engineering is legal if the information is obtained through legal means, but one must be careful with the information that is obtained and not copy it unless legally allowed to do so. The Digital Millennium Copyright Act of 2000 prohibits breaking technological protections, such as encryption, to learn about a competitor's product.

Copyrights protect published works such as books, articles, music, and software. A copyright means that others cannot distribute copyrighted material without permission of the owner. It is easy to obtain a copyright—all that a person has to do to copyright material is indicate the word "copyright" on the work, along with the year of publication, and identify the name of the copyright holder. Copyrights can be officially registered through the

U.S. Copyright Office, and it is a good idea to do so for involved works such as writing a book or publishing music. Registering provides a stronger legal basis for pursuing claims. Copyrights are good for the lifetime of the holder plus 50 years, while for a company they are good for 75 years.

11.3.2 Liability and Negligence

Civil lawsuits, which are those that are relevant for our discussion, are those in which one party sues another. They involve something that is known as a *tort*, which serves as grounds for the lawsuit. A tort is a wrongful act, though not necessary illegal, for which a civil lawsuit can be brought, including product liability. A company or person can be sued for damages caused by a product design and be held *liable* for them—meaning required to pay monetary damages.

The two standards for determining legal liability in tort law are that of negligence and strict liability. In the case of *negligence*, it must be shown that the manufacturer did not follow reasonable standards and rules that apply to the situation and also committed a wrongful act. Exactly what constitutes reasonable has to be determined for the particular case. For a product design, a manufacturer can be held legally liable for negligence if the plaintiff demonstrates that the following four conditions hold true:

1. The manufacturer had a duty to follow reasonable standards and rules.
2. There was a breach of duty (i.e., failed to include safety devices).
3. The plaintiff was harmed.
4. The breach caused the harm.

Depending on the severity of the danger, there are different levels of negligence: simple, gross, and criminal. Negligence claims can be brought for design flaws, manufacturing defects, and for failing to warn the user of safety hazards.

An even less stringent standard than negligence is known as **strict liability**, in which the following four conditions must hold true:

1. The product was dangerous and/or defective.
2. The defect existed when it left the manufacturer's control.
3. The defect caused harm.
4. The harm is assignable to the defect.

Strict liability focuses only on the product itself—if the product contains a defect that caused harm, the manufacturer is liable. This is regardless of whether there was negligence, if safety devices were incorporated, or if the user was warned of potential dangers. If the product had a defect or was dangerous when it left the hands of the manufacturer, the manufacturer may be liable.

11.4 Handling Ethical Dilemmas

You are going to encounter ethical dilemmas in your career, both technical and based upon interpersonal relations. Ethical dilemmas are not always obvious and can be quite subtle. A supervisor probably won't say "Could you falsify this data for me so we can ship the product?" The pressure will likely be much more subtle and over the course of multiple conversations, unlike Dilbert's dilemma in Figure 11.2. Consider the following sequence of statements:

- "We have invested a lot of time and money in the design."
- "We really need this system to work."
- "The company's future depends upon this."
- "Is there any way that we can make adjustments to make it pass the certification?"
- "Is it close enough that we could certify it? It really meets the needs and the standard has a margin of error built in to it."

Figure 11.2 Dilbert's ethical dilemma. (Dilbert © Scott Adams / Dist. by United Feature Syndicate, Inc.)

A framework for considering decisions is shown in the 2 × 2 matrix shown in Figure 11.3 [Die00]. The idea is that decisions have both ethical and legal dimensions to them. The legality may be in terms of internal company policies or local, state, and federal laws. Quadrant I decisions are clearly to be avoided, as they are not legal or ethical. Quadrant II decisions present an interesting dilemma since they are legal, but yet are not ethical. Making such a decision may not have punitive ramifications, but will have a negative impact on your professional reputation. Quadrant III decisions certainly feel right and are tempting, since they are moral but not legal. It is in Quadrants II and III where most moral dilemmas take place. Taken together, II and III represent opportunities for reform, to change the system positively so that ethical choices are legal and unethical choices are illegal. Left uncorrected, both lead to cynicism with the system, whether it is company policy or the legal system. Correcting them may take longer to address than the immediate dilemma, but have the potential for high payoff. Clearly, quadrant IV decisions are best and the goal in the decision process.

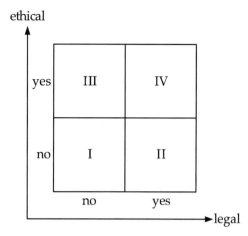

Figure 11.3 A 2 × 2 matrix for ethical decision making [Die00].

One way to evaluate a decision is through what is known as the *newspaper test*. The idea is to ask whether you would be comfortable if the decision were published in a newspaper for all to see. Advice from trusted friends or colleagues can be valuable, but if they are affected by the decision, they may not be a good source. Many companies have an ethics office that provides guidance to which employees can turn. The IEEE and Online Ethics Center for Engineering and Science co-sponsor an ethics help line at *www.OnlineEthics.org*, where ethical questions can be submitted.

There may be situations in which you are put in an ethical dilemma by your employer that cannot be resolved internally. As a last resort, you may go outside of the company to the press or a government agency to report the problem. This is known as being a *whistleblower*. An example of this is the space shuttle Challenger accident, in which the explosion was caused by O-ring failures in the booster rockets made by Morton-Thiokol. The engineers and scientists recommended against the launch, but their superiors at NASA and Morton-Thiokol went ahead over their objections. The engineers later "blew the whistle" by going public with this information. Although it was too late to avert the disaster, it was valuable in revealing what had happened and preventing future mishaps. The following criteria should be satisfied when you are considering becoming a whistleblower [Deg81]:

- The harm to the public must be considerable or serious.
- Concerns must have been expressed to your superiors (up to the CEO) without satisfactory resolution.
- You have documented evidence that would convince an impartial observer that your company is wrong. You should have clear technical information to support your claims.
- Release of the information outside of the company will prevent the harm.

Whistleblowing does have risks, particularly loss of job, but the risks may be well worth it. The Sarbanes-Oxley Act of 2002 improved federal protection of whistleblowers in the wake of corporate scandals in the 1990s.

11.5 Case Study Analysis

A good way to develop decision-making skills is by examining case studies. We now go through the steps of analyzing a case study, employing the IEEE Code of Ethics as a guide. In order to do so, we apply a paradigm that is a modification of the one used by Lockheed Martin Corporation in their employee ethics training programs [Loc97]. The modified paradigm is as follows:

1. *Gather information.* What things are known about the situation, but also very important, what isn't known, and what assumptions are being made?
2. *Identify the stakeholders.* Who will be affected by the decision? They may include you, your company, your supervisors, your professional colleagues, your profession, the public, and users of your products.
3. *Consider what ethical values are relevant to this situation.* Identify the elements of the IEEE Code of Ethics and legal issues that apply to the situation.
4. *Determine a course of action.* Identify different alternative decisions and actions. Select the one that you believe best meets the interests of the stakeholders and the ethical values.

Example 11.1 demonstrates the application of this paradigm to a case study for a flawed chip design. More case studies are provided in the end-of-chapter problems.

Example 11.1 Ethics Case Study for a Flawed Chip Design. (Texas A&M Ethics Case Studies, *http://ethics.tamu.edu.* Reprinted by permission.)

Your company has designed a chip for a new scientific calculator that features high-precision floating point accuracy to 17 significant digits for all 250 mathematical functions provided. After 1½ years in development, and after shipping over 500 beta units to key customers, the company discovers a problem with certain calculations. In addition, the company has manufactured 5000 more calculators with this chip that are ready to be shipped.

In order to expedite floating point operations (used in handling scientific notation in mathematical operations) in a computer or calculator, certain tables of values often are used to assist in the speed of execution. (For example, a calculator requiring as long as 3 minutes to perform a tangent calculation would have no market appeal.) These tables can contain up to 100 integer entries. During beta testing, you discover that several of these values were incorrectly entered before they were burned into the firmware.

Further testing concludes that because of the location and use of these table errors, the only mathematical results affected will occur in the 13th to the 17th significant digits for the double-precision floating point operations. Your management is applying subtle pressure to release the chips because of the time and money invested in the project so far. Identify a plan of action.

Discussion:

Step 1: Gather information

The information makes it clear that the calculator may fail if precision beyond the 13th digit is needed. What is not known is the standard precision of a calculator. An Internet search of calculator specifications was conducted, and it was found that most scientific calculators have a 10- or 12-digit display plus two digits for the power. Going to the 13th digit implies a higher than normal precision calculator. A calculator with 17 significant digits would likely be used by scientists and engineers for high-precision calculations.

Step 2: Identify stakeholders

The following are possible stakeholders:

- Users of the calculator. In certain situations, the user can make an incorrect calculation.
- Public. The users of calculator may perform calculations that could affect safety of public. This could be a real possibility given the likely users.
- Company and employees. There are negative ramifications of releasing a faulty product. This could result in monetary harm to the employees of the company that you work for.

Step 3: Identify relevant ethical values

Values from the IEEE Code are identified with a discussion of why they apply:

1: Release of the faulty calculator has the potential to endanger the safety of the public.

3: There is a need to be honest in stating claims as to the precision of the calculator.

9: There is clear potential to injure the reputation of the company and its employees.

In terms of legal issues, the company would be opening itself to claims of negligence, particularly since the defect was identified prior to release of the product.

Step 4: Determine a course of action

Three possible courses of action are:

A: Release the product as is without notifying the customers. This is not a good choice because of the potential harm to the safety of the public. This would not pass the newspaper test. It also may be illegal if the calculator is advertised to work to 17 significant digits.

B: Use the chips in a different calculator that is guaranteed to compute only at a lower precision, if such an option exists. The company would have to be producing one, and the technology would have to be compatible.

C: Throw away the chips and take a loss on their production and correct the problem. Conduct testing again to verify that the corrected chips work properly.

Options B and C are both reasonable choices. Option B reduces economic losses, but acceptance testing should be conducted to make sure they operate properly at the lower precision. Option C is the safest solution. Option A might be seen as a viable choice, however, since it could be reasoned that handheld calculators are not often used for safety-critical applications. That is a risky assumption and the potential negative effects are great. Also note that the ideal situation would have been to develop a test plan (Chapter 7) that would have caught this error before release.

11.6 Project Application: Incorporating Ethics in the Design Process

There are good design practices that can be followed to minimize the chances that a product will be unsafe. The first and most important issue is to identify safety and health factors that impact the system and address them in the design. Its importance is reflected clearly as the first item in the IEEE Code of Ethics. It is a fundamental canon in virtually all professional engineering ethical codes. Be aware that there are often tradeoffs between economic and safety constraints and one may be satisfied at the expense of the other. It is also important to follow good practices throughout the design process. Ways that ethical considerations are included throughout the design process are:

- *Conduct research so that prior art is understood.* This will minimize conflicts over intellectual property ownership.
- *Make sure that the requirements meet the needs of the stakeholders.* If they do not, the wrong system will be designed. This is the concept of validation that was addressed in Chapter 3. Be honest and realistic in making claims about what the system can do and realize that all have limitations. There is often tension between engineering and marketing functions of an organization, with engineers often believing that marketing is overstating the capabilities of the product. A properly developed requirements specification is an effective tool for communicating the proposed system functionality to all parties.

- *Identify and apply safety standards.* There are many codes and guidelines that address safety in design for particular technologies. They should be identified and applied as necessary. For example, it may be wise to design a consumer product to meet UL (Underwriters Laboratories) guidelines. It may difficult to fully incorporate guidelines in a capstone design project, depending upon the complexity of the design and the standards. However, consideration should be given to identifying those that apply.

- *Keep the design space large and explore as many solutions as possible.* The hallmark of design is that there is typically no single solution to the problem, but many that may **satisfice**. Satisfice means that a solution may meet the design requirements, but may not be the optimal solution. It is important to fully understand the technical tradeoffs involved in a given design and their impact on health and safety.

- *Consider all of the possible ways that a system can fail.* This follows the IEEE Code of Ethics, which indicates that engineers should understand technology, its application, and potential consequences. This can be done formally through the use of techniques such as failure mode analysis and the examination of system reliability (Chapter 8). In safety critical systems, the use of redundant systems may be warranted to improve reliability. An excellent example of a technical design that incorporates redundancy in a safety critical application for a hot tub controller is found in the article "Designing for Reliability, Maintainability, and Safety" by George Novacek in Circuit Cellar magazine [Nov00, Nov01].

- *Identify ways in which the system can fail by misuse or operator error.* This requires thinking like a beginner who is completely unfamiliar with the technology. Provide manuals for operation and safety labels where appropriate.

- *Make realistic cost and project schedule estimates.* The IEEE Code of Ethics indicates that engineers should be honest and realistic in stating claims, and this applies to the project plan as well as the system cost.

- *Conduct design reviews.* The purpose of a design review is to have peers who are knowledgeable about the design and the technology conduct a review of the work you have done. This can be a humbling, but valuable, experience. It is also in keeping with the IEEE Code of Ethics, which indicates that engineers should seek, accept, and offer honest criticism of technical work and acknowledge and correct errors.

- *Verify the engineering requirements during testing.* We saw in Chapter 3 that verification is the process of showing that the system is built properly. Verification usually occurs at the end of a project at a time that there may be a great deal of pressure to complete the tests and finish the project.

11.7 Summary and Further Reading

This chapter provided a brief overview of ethics, morality, and different ethical decision-making systems. The IEEE Code of Ethics was presented, which serves as a set of ethical values for the electrical and computer engineering profession and provides a basis for analyzing ethical dilemmas. Legal issues surrounding intellectual property and product liability were provided. Advice for incorporating safety and ethics in design projects was provided as well as a paradigm for analyzing case studies.

The background on ethical frameworks and its importance in engineering came from a number of articles. The article "What Has Ethics to Do with Me?: I Am an Engineer" [Gob99] provides background on rule-based ethics, universality, and transitivity. "A Piercian Approach to Professional Ethics Instruction" [Cha02b] addresses ethical decision-making frameworks. "Ethical Considerations in the Engineering Design Process" [Van01] emphasizes the need to keep the design space as large as possible. The article "Integrating Ethics and Design" [Mcl93] does a good job of describing three different levels of ethical decision making in technical, professional, and social contexts. The two engineering design textbooks by Dieter [Die00] and Hyman [Hym98] are good resources on legal and intellectual property issues. The United States Patent and Trademark Office has an excellent website (*www.uspto.gov*) with information on patents, applying for a patent, and a searchable patent database. The website *www.HowStuffWorks.com* is also a good source on ethical theory, patents, and legal issues surrounding product liability.

11.8 Problems

11.1 Describe the relationship between ethics and morals.

11.2 Describe the differences between morals and values.

11.3 Which patent is most relevant for engineering inventions, a design patent or utility patent? Why?

11.4 What are the criteria that are used in evaluating patents?

11.5 Explain the importance of claims in a patent application.

11.6 Discuss the tradeoffs involved between using patents and trade secrets to protect intellectual property.

11.7 When can reverse engineering be used, and how can the information obtained from it be used?

11.8 What is the difference between negligence and strict liability in tort law?

11.9 For the case study presented below, apply the ethical decision-making paradigm presented in Section 11.5 to analyze the situation. Present potential solutions to the scenario and provide a discussion of them.

Case Study: Disk Drive Diagnostics. (Copyright John Wallberg. Reprinted by permission.)

SCSI, an industry standard system for connecting devices (like disks) to computers, provides a vendor ID protocol by which the computer can identify the supplier (and model) of every attached disk.

Company C makes file servers consisting of a processor and disks. Disks sold by C identify C in their vendor ID. Disks from other manufacturers can be connected to C's file servers; however, the file server software performs certain maintenance functions, notably prefailure warnings based on performance monitoring, only on C-supplied disks.

Company P decides to compete with C by supplying cheaper disks for C's file server. They quickly discover that while their disks work on C's file servers, their disks lack a prefailure warning feature that C's disks have. Therefore, the CEO of P directs you, the engineer in charge of the disk product, to find a solution to the problem of no prefailure warning for your disks. Using reverse engineering, you discover that by changing the vendor ID of P's disks, the C file servers will treat P disks as C disks. Your management at company P instructs you to incorporate this change into your product so that you can advertise the disks as "100% C-compatible." What would you do in this situation?

11.10 For the case study presented below, apply the ethical decision-making paradigm presented in Section 11.5 to analyze the situation. Present potential courses of action and provide a discussion of them.

Case Study: Encryption Software (Texas A&M Ethics Case Studies, *http://ethics.tamu.edu*. Reprinted by permission.)

You are a recently hired engineer who has been recruited directly out of college. For your first assignment, your boss asked you to write a piece of software to provide security from "prying eyes" over emailed documents; these documents would be used internally by the company. This software will subsequently be distributed to different departments.

Upon completion of this software project, you saw a program on the local news about an individual in California who has made similar software available overseas. This individual is currently under prosecution in a federal court for the distribution of algorithms and information which (by law) must remain within the United States for purposes of national security.

It occurs to you that your company is a multinational corporation and that the software might have been distributed overseas. You then discover that the software has indeed been sent overseas to other offices within the corporation. You speak with your boss, informing him of the news program from the night before. He shrugs off this comment, stating that "The company is based in the United States and we are certainly no threat to national security in any way. Besides, there's no way anyone will find out about software we use internally."

You agree with your boss, and let it go. Later on however, you receive a letter from someone working as a contractor for the company overseas. Through some correspondence regarding the functionality of the software and technical matters, you learn that the Middle Eastern office had been supplying this software outside the company to contractors and clients so that they could exchange secure emailed documents. What would you do in this situation?

11.11 For the case study presented below, apply the ethical decision-making paradigm presented in Section 11.5 to analyze the situation. Present potential courses of action and provide a discussion of them.

Case Study: A Failure. (Texas A&M Ethics Case Studies, *http://ethics.tamu.edu*. Reprinted by permission.)

You work for Velky Measurement, which has for years provided DGC Corporation with sophisticated electronic equipment for patient health monitoring systems. Recently, DGC returned a failed piece of measurement equipment. A meeting was held with representatives of Velky and DGC to discuss the problem. This included you and your project manager who is intimately acquainted with the returned equipment. During the course of the meeting it becomes apparent to you that the problem has to be Velky's. You suspect that the equipment failed because of an internal design problem and that it was not properly tested. However, at the conclusion of the meeting your project manager represents Velky's official position—the test equipment is functioning properly.

You keep silent during the meeting, but afterwards talk to your project manager about his diagnosis. You suggest that Velky tell DGC that the problem is due to a design fault and that Velky will replace the defective equipment. Your manager replies, "I don't think it's wise to acknowledge that it's our fault. There's no need to hang out our wash and lessen DGC's confidence in the quality of our work. A good will gesture to replace the equipment should suffice."

Utlimately, Velky's management replaces the equipment because DGC has been such a good customer. Although Velky replaces the equipment at its own expense, it does not disclose the real nature of the problem. What would you do in this situation?

11.12 For the case study presented below, apply the ethical decision-making paradigm presented in Section 11.5 to analyze the situation. Present potential courses of action and provide a discussion of them.

<u>Case Study: A Vacation</u> (Texas A&M Ethics Case Studies, *http://ethics.tamu.edu*. Reprinted by permission.)

You work for Rancott and were looking forward to an upcoming trip for weeks. Once you were assigned to help install Rancott's equipment for Boulding Corporation, you arranged a vacation at a nearby ski resort. The installation was scheduled to be completed on the 12th and your vacation would begin on the 13th. That meant a full week of skiing with three of your old college buddies.

Unfortunately, not all of the equipment arrived on time. Eight of the ten identical units were installed by midmorning on the 12th. Even if the remaining two units had arrived that morning, it would take another full day to install them. However, you were informed that it might take as long as two more days for the units to arrive.

"Terrific," you sighed, "there goes my vacation—and all the money I put down for the condo." "No problem," replied Jerry, the Boulding engineer who had worked side-by-side with you as each of the first eight units was installed. He said "I can handle this for you. We did the first eight together. It's silly for you to have to hang around and blow your vacation." Jerry knew why you were sent to supervise the installation of the new equipment. It had to be properly installed in order to avoid risking injuries to those who use it. Although you are aware of this, you are confident that Jerry is fully capable to supervise the installation of the remaining two units. What would you do?

Chapter 12 Oral Presentations

Nothing should be explained in a way that it cannot be understood by an intelligent 12 year old. — Albert Einstein

We can all probably remember the anticipation of our first oral presentation—the sweaty palms, butterflies in the stomach, and the pressure of trying to remember all the points to be made. Then there is the fear associated with standing up in front of peers or teachers and presenting ideas to have them openly criticized. According to a 1973 London Times survey, Americans are more afraid of speaking in front of groups than dying. Perhaps this is due to the fact that, although people know they will die someday, the danger associated with giving a presentation is more imminent and a greater concern. Somewhere in the capstone design experience, it is likely that you will have to make an oral presentation. Examples are the project proposal, a midterm design review, and the final presentation. The ability to communicate your ideas is important beyond your academic career, since practicing engineers are often required to make oral presentations. Further, your overall ability to communicate influences how others will accept your ideas and act upon them—those who communicate clearly are held in high regard by their peers and tend to advance more quickly. The good news is that there is help to overcome the fear of oral presentations. With practice and adherence to some basic principles, one can become a competent, if not excellent, communicator.

Learning Objectives

By the end of this chapter, the reader should:
- Understand how people evaluate oral presentations.
- Understand common elements of a technical presentation.
- Be able to assemble an effective presentation.

DILBERT® by Scott Adams

Figure 12.1 PowerPoint™ disability. (Dilbert © Scott Adams / Dist. by United Feature Syndicate, Inc.)

12.1 How People Evaluate Presentations

It is informative to understand how your audience responds to and evaluates oral presentations. Listening to a presentation is strongly associated with what is referred to as right-brain activity. Right-brain activity is dominated by emotion and intuition, while left-brain activity is associated with logical thinking and reason. This is an oversimplified model of the brain, but the point is that emotion and intuition are important elements that people rely on when evaluating a presentation.

There are three elements, known as the "three V's," that constitute a presentation: the verbal, the vocal, and the visual. Verbal is what the speaker says—the actual words and content that come out of the speaker's mouth. Vocal is indicative of how it is said, and includes pitch, enthusiasm, inflection, and intonation. Visual is what the audience sees—the speaker's appearance, eye contact, posture, facial expressions, and gestures. All three factors go into the evaluation of speakers, but what is the relative importance of each? The results of a 1964 UCLA study by Dr. Albert Mehrabian (who has bachelor's and master's degrees in engineering and a Ph.D. in psychology) indicates that the impact of the three elements is 7% verbal, 38% vocal, and 55% visual. That seems disappointing because we would like to think that the content of the presentation is most important. Realize that the numbers come from a simplified study and it is likely that the percentages would be different for a highly technical audience. The point is that content is important, but the other elements can't be ignored. If the visual and vocal aspects of the presentation are poor, it will be perceived negatively and make it difficult for the audience to accept and pay attention to the information presented.

Here is another consideration to think about the next time you make a presentation or meet somebody. In the first 7 seconds of meeting someone, people typically form a great number of subconscious opinions about the person they meet [Bai81]. This includes the person's income level, education level, competence, character, trustworthiness, personality, confidence, intelligence, work ethic, and dependability. What factors are the opinions based on? They include appearance, dress, posture, and speech patterns.

12.2 Preparing the Presentation

In order to make an effective presentation, the presenter must understand the subject matter (substance counts), understand the needs of the audience, and prepare the presentation. The remainder of this section provides guidance for preparing the presentation.

Analyze the Audience

An oral presentation is for the benefit of the audience, not the presenter. It is necessary to analyze and understand the audience's needs and prepare the presentation to meet them. For example, a presentation for engineering professors would likely be different from one for your family and friends. Analyzing the audience is no different than the process that one goes through when writing a document. Some questions to ask in this process are [Bai81]:

- What are they interested in?
- What do they want from your talk?
- What does the audience already know about my subject?
- What don't they know?
- What is the attitude of the audience toward me and my subject?
- What are the values of the audience?
- What do you want them to know or learn?

Understanding the needs of the audience and putting their interests first establishes credibility so that they are more willing to accept the content of the presentation.

Before creating the presentation, identify the main points that the audience should take away from it. A rule of thumb is to identify three main points for a talk, as people tend to forget more than that. Although it is not a strict rule, keep the number of points in that range, say two to five. Once the points are identified, organize the presentation to support them.

Organize the Presentation

Just like a story, a presentation has an introduction, a body, and a conclusion. This is encapsulated in the often-heard wisdom for presentations to "tell them what you are going to tell them" (the introduction), "tell it to them" (the body), and "tell them what you just told them" (the conclusion).

The introduction is absolutely critical—if the audience does not understand the presentation from the outset, they will tune out. Einstein's advice that nothing should be explained in such a way that it cannot be understood by an intelligent 12-year-old is particularly relevant here. Take time to explain the problem in simple terms. Part of an effective introduction is obtaining the interest of the audience. There are many ways to accomplish this, and examples include the use of rhetorical questions or the narration of an experience that the audience can relate to. This should have to do in some way with the information being presented. Overall, the objective is to motivate the audience by describing what is being presented and why it is important. After giving motivation of the problem, an overview of the talk can be provided. The overview should be relevant to the problem at hand, not a generic one that can apply to virtually any presentation.

Structure the body of the presentation to support the main points. This is done by having a group of two to four related slides that support each of the main points. The first slide of the group provides some key ideas, followed by the remaining slides that go into more detail on the particular point. Don't make the talk unnecessarily technical or use a lot of jargon. This does not mean that it should not have technical content, but that judgment should be exercised in presenting the right amount of detail. The level of technical detail depends upon the education and experience of the audience. If it is necessary to use jargon or acronyms, make sure that they are defined for the audience. Consider alternative ways of explaining things. The use of analogies is particularly powerful when explaining complex and abstract material. One strategy is to increase the level of complexity as the talk proceeds. That way, much of the early material will be understandable to the majority of the audience, while the latter more complex material may be understood by only a small fraction. Everybody will then leave the presentation with some understanding of the content.

In a typical engineering classroom lecture, the professor usually goes through many steps in defining and deriving equations. Realize that the goals of a classroom lecture are very different from that of an oral presentation. When working with equations, don't derive or give too many intermediate steps, unless that is the point of the presentation. Provide assumptions, selected intermediate equations, and the important results. Audiences generally assume you have done your homework and derived the equations properly. There is a tendency to present equations, vaguely refer to them, and then move on. Equations should be presented for a reason, so talk about them and describe their significance. Every equation has its own story; it is the presenter's job to tell it. The same is also true of graphs and plots.

The conclusion provides the opportunity to summarize and emphasize the main points of the presentation. Again, tell them what you told them by reviewing the important points and conclusions. That way if somebody was lost during the presentation, they can understand the importance of the work. If there are recommendations to be made for future action, address them here. The conclusion is also an opportunity to explain the next steps for the project.

Lay out the Slides

Below are pointers for the layout of the slide content.

- *Use a large font.* This ensures that information on the slides can easily be seen by the audience; 24-point or greater font is typically sufficient.

- *Have a goal of five to seven bullet points per page.* Avoid the tendency to cram as much information as possible on a page, which is often done so that the presenter does not neglect any material. Avoid this, and use five to seven bullet points to guide the discussion. Presentation software packages, such as Microsoft PowerPoint, allow you to introduce bullet items one at a time, which helps to keep the discussion on track.
- *Avoid fancy graphics and special effects that add no value.* Presentation packages allow the addition of fancy features, such as spiraling text and sound effects. They add little to the presentation and when overused distract from the content. The content and material are what matter the most, not fancy formatting and special effects. To quote Edward Tufet, professor emeritus at Yale,

    Power Corrupts, PowerPoint Corrupts Absolutely.
- His point is that fancy graphics and features are used far too much with PowerPoint presentation software, and that this overpowers both the content and the audience. [Tuf03].
- *Group slides together to make a major point.* Make the first slide the general one with key statements. The following ones should have more detailed information supporting the point.
- *Do not create a canned talk or speech.* That is acceptable in some fields, but not in engineering and science where a more extemporaneous style is the norm. Let the bullet points and other material on the slides serve as guides for what to say. Avoid the use of cue cards and do not just read directly from the slides for the presentation.

Meet the Time Constraints

Make sure that the presentation falls within the time constraints—the audience will be alienated if it is far too short or too long. The tendency is to exceed the time limit since there is so much information that the presenter wants to convey. You may be abruptly cut off and not be able to conclude the presentation if the time limit is exceeded. Think about this—how would you describe all that you know about electrical or computer engineering in 10 minutes? It is challenging, but if you had only 10 minutes you would probably give a brief overview of the major accomplishments made in the field. Accept that all of the information can't be conveyed in the given time and use it carefully to highlight the important material.

A heuristic is to take the length of the allotted time in minutes and divide it by two. That provides an estimate of the number of slides to prepare. Once the presentation is prepared, practice to see if it can be presented reasonably in the time allotted. Practice the talk in front of your teammates, boyfriend, girlfriend, mother, or pet rock. Be careful not to overprepare to the point of sounded scripted. Practice the talk the night before the presentation and do only a brief review of the material right before the talk. Be sure to allow time for the question-and-answer session that usually occurs at the end of a presentation.

Prepare for the Question and Answer Session

One of the biggest fears of presenters is the dreaded question and answer session. This is where the audience gets to ask questions and possibly expose the presenter for what he or she doesn't know. For example, the questioner may ask "Are you familiar with the work of Johnson and Smith from 1984 in which they proposed exactly the same idea as yours?"

How do you prepare for this? You must be knowledgeable about the subject, but you don't need to have the answers to every possible question. It is good practice to rephrase questions that are asked for the benefit of you, the audience, and the questioner. Rephrasing the question ensures that you are answering the correct question (how many times have you been annoyed when a teacher answered the wrong question?) and provides time to think and formulate a response. It demonstrates to the questioner that you understand the question and are able to present it in a different format. If the questioner is hostile, make sure that you rephrase the question in a positive light. Rephrasing is also a courtesy for the other members of the audience who may have not heard or understood the question.

Most questions are made in good faith as the questioner is trying to clarify a point or learn more. Sometimes, questioning can become hostile or aggressive. If this happens, make sure not to respond in kind or put the questioner down. The presenter has the position of power and becomes a bully by doing this. Be sure to maintain eye contact with the questioner, smile, and remain relaxed in your responses. If you can't answer the question, admit it and don't try to come up with a phony answer. If the questioner is persistent, offer to discuss it in more detail with them after the presentation.

12.3 Project Application: Design Presentations

Examples of the three presentations that you may make during the design process are listed in Table 12.1. It is a guide of points to consider preparing presentations and should be adjusted to meet particular needs of the situation. The checklist in Table 12.2 is provided to aid in preparing for presentations.

Table 12.1 Guide for preparing design presentations. The chapters associated with the points are identified.

Presentation	Points to Consider in Preparing the Presentation
Project Proposal	*Introduction.* Provide an overview of the project and address the need, motivation, goals, and objectives. The audience is probably not familiar with the concept and it is important to describe the problem in simple and concise terms (Chapter 2).
	Problem Analysis. Indicate what the current state of the art in the field is regarding the technology. If it is a new product concept, identify similar products that are available and what is unique about this one. If it is a research-oriented project, include the basic theory and address current status of work in this area (Chapter 2).
	Requirements Specifications. Address the engineering requirements and provide a justification for their selection. Describe the standards and constraints that apply to the problem (Chapter 3).
	Preliminary Design Options. Depending upon progress, some preliminary options for the design may have been developed and can be presented here (Chapter 4).
Design Review	*Introduction.* Provide a brief overview of the motivation for the project.
	Requirements. Recap the critical requirements that have to be met.
	The Proposed Design. Present the high-level design. Explain how it works and how the pieces fit together. Include design details of the subcomponents and systems. Address how the proposed design meets the engineering requirements. Identify the alternatives investigated (Chapters 4, 5, and 6).
	Preliminary Test Results. Include test and prototype results (Chapter 7).
	Project Plan. By this point (if not earlier), a project plan should be in place, so consider presenting a summary of the plan (schedule, responsibilities, and cost) (Chapter 10).
Final Presentation	*Introduction.* Provide an overview or motivation for the project.
	The Final Design. Describe the final design implementation. A good way to organize is to provide a high-level overview of the design and describe how it operates. Then, provide detail and a description for each of the successive hardware/software subsystems (Chapters 5 and 6).
	Testing and Results. Describe/demonstrate the key tests and results that show the functionality of the design. Provide demonstrations if appropriate. Indicate how the final realization did or did not meet the requirements (Chapter 7).
	Conclusions. Summarize conclusions about the project and provide recommendations for further work. Indicate lessons learned.

Table 12.2 Checklist and self-assessment for oral presentation preparation. Score the elements as 1 = strongly disagree, 2 = disagree, 3 = neutral, 4 = agree, 5 = strongly agree.

Organization	Score
The background and needs of the audience were analyzed.	
The main points of the presentation are identified.	
The motivation is clear and would be understandable to an intelligent 12-year-old.	
An overview of the presentation is provided. It is relevant to the presentation, not a generic one that can be used in any presentation.	
The body is organized to support the main points.	
The conclusion summarizes the main points and future work.	
Visual Aids	
The fonts and graphics are large enough to be seen by the audience.	
The equations are of the right number and level. The presenters are prepared to discuss any equations presented.	
The slides are arranged to support the main points.	
The presentation does not contain unnecessary graphics and special effects.	
Presentation Delivery	
The presentation has been rehearsed. It meets the time constraints and there is sufficient time for questions and answers.	
Voice projection is loud enough so that the audience can hear the presenters.	
All members participate in the presentation and have reasonably equal responsibilities. (If one team member always presents the introduction and another the technical material, it is a sign that not all members are participating equally on the project.)	
The presenters do not rely on cue cards.	
The presenters are comfortable in front of an audience. (Do they make good eye contact with the audience? Do the presenters move around the room or do they stand stiffly behind a podium?)	
All presenters are knowledgeable on the subject and prepared to answer questions.	
The presentation software was tested on the platform to make sure it works.	
The presenters are dressed properly for the occasion.	

12.4 Summary and Further Reading

During the design process and your professional career, you will need to communicate ideas effectively. One of the most common ways to do this is via an oral presentation. Visual, verbal, and vocal aspects influence the effectiveness of an oral presentation. Although the verbal aspect, or content, is important, the visual and vocal delivery aspects strongly affect the audience's perception and cannot be overlooked. In preparing the presentation, the needs and background of the audience should be taken into consideration and the main points to be conveyed identified. Creating the presentation is much like telling a story—there should be an introduction, a body, and a conclusion that are organized to support the main points. The slides should be grouped to support these points. Concepts should be explained as simply and clearly as possible. Increasing the complexity as the presentation proceeds will allow the presentation to reach all members of the audience to some extent. Practicing the presentation is especially important for novice presenters, and tips were provided for meeting the time constraints and preparing for question sessions.

There are many excellent resources and articles available regarding oral presentations. A concise and humorous article geared for new speakers in the technical fields is "Advice to Beginning Physics Speakers" by James Garland [Gar91]. The IEEE Transactions on Professional Communications journal addresses many aspects of communications including oral presentations. Mindtools (*www.mindtools.com*) has a section on communication skills that includes a preparation checklist on presentation, delivery, appearance, and visual aids. "A Good Speech Is Worth a Thousand Words" by Bert Decker [Dec84] addresses right and left brain thinking as well as the three V's of giving a presentation. The article "How to Overcome Errors in Public Speaking" by John Baird [Bai81] addresses how to analyze the audience, the judgments that are made when meeting a person, the introduction, and conclusion. Other resources used in the preparation of this chapter include: "The Engineering Presentation—Some Ideas on How to Approach and Present It" [Ros93], "Handling a Hostile Audience—With Your Eyes" [Car89], and "How to Speak so Facts Come Alive" [Ste89]. Many of the references are compiled in the book Writing and Speaking in the Technology Professions: A Practical Guide edited by David Beer [Bee03].

Appendix A Glossary

acceptance test An acceptance test verifies that the system meets the **requirements specification** and stipulates the conditions under which the customer will accept the system (Chapter 7).

activity An activity is a combination of a **task** and its associated **deliverables** that is part of a project plan (Chapter 10).

activity on node A form of a **network diagram** used in a project plan. In the activity on node (AON) form, activities are represented by nodes and the dependencies by arrows (Chapter 10).

activity view The activity view is part of the **Unified Modeling Language**. It is characterized by an activity diagram; its **intention** is to describe the sequencing of processes required to complete a task (Chapter 6).

analytical hierarchy process (AHP) A decision-making process that combines both quantitative and qualitative inputs. It is characterized by weighted criteria against which the decision is made, a numeric ranking of alternatives, and computation of a numerical score for each alternative (Appendix B and Chapters 2 and 4).

artifact System, component, or process that is the end result of a design (Chapter 2).

automated script test An automated script test is a sequence of commands given to a unit under test. For example, a test may consist of a sequence of inputs that are provided to the unit, where the outputs for each input are then verified against prespecified values (Chapter 7).

baseline requirements The original set of requirements that are developed for a system (Chapter 3).

black box test A test that is performed without any knowledge of internal workings of the unit under test (Chapter 7).

Bohrbug Bohrbugs are reliable **bugs**, in which the error is always in the same place. This is analogous to the electrons in the Bohr atomic model, which assume a definite orbit (Chapter 7).

bottom-up design An approach to system design where the designer starts with basic components and synthesizes them to achieve the design objectives. This is contrasted to **top-down** design (Chapter 5).

brainstorming A free-form approach to concept generation that is often done in groups. This process employs five basic rules: (1) no criticism of ideas, (2) wild ideas are encouraged, (3) quantity is stressed over quality, (4) build upon the ideas of others, and (5) all ideas are recorded (Chapter 4).

brainwriting	A variation of **brainstorming** where a group of people systematically generate ideas and write them down. Ideas are then passed to other team members who must build upon them.
break-even point	The break-even point is the point where the number of units sold is such that there is no profit or loss. It is determined from the total costs and revenue (Chapter 10).
bug	A problem or error in a system that causes it to operate incorrectly (Chapter 7).
cardinality ratio	The cardinality ratio describes the multiplicity of the entities in a relationship. It is applied to **entity relationship diagrams** and Unified Modeling Language **static view diagrams** (Chapter 6).
class	Classes are used in object-oriented system design. A class defines the attributes and methods (functions) of an **object** (Chapter 6).
cohesion	Refers to how focused a module is—highly cohesive systems do one or a few things very well. Also see **coupling** (Chapter 5).
concept fan	A graphical tree representation of design decisions and potential solutions to a problem. Also see **concept table** (Chapters 1 and 4).
concept generation	A phase in the **design process** where many potential solutions to solve the problem are identified (Chapter 1).
concept table	A tool for generating concepts to solve a problem. It allows systematic examination of different combinations, arrangements, and substitutions of different elements for a system. Also see **concept fan** (Chapter 4).
conditional rule-based ethics	An ethics system in which there are certain conditions under which an individual can break a rule. This is generally because it is believed that the moral good of the situation outweighs the rule. Also see **rule-based ethics** (Chapter 11).
constraint	A special type of requirement that encapsulates a design decision imposed by the environment or a stakeholder. Constraints often violate the abstractness property of engineering requirements (Chapter 3).
controllability	A principle that applies to testing. Controllability is the ability to set any node of the system to a prescribed value (Chapter 7).
copyright	Copyrights protect published works such as books, articles, music, and software. A copyright means that others cannot distribute copyrighted material without permission of the owner (Chapter 11).
coupling	Modules are coupled if they depend upon each other in some way to operate properly. Coupling is the extent to which modules or subsystems are connected. See also **cohesion** (Chapter 5).

creative design	A formal categorization of design projects. Creative designs represent new and innovative designs (Chapter 2).
critical path	The path with the longest duration in a project plan. It represents the minimum time required to complete the project (Chapter 10).
cross-functional team	Cross-functional teams are those that are composed of people from different organizational functions, such as engineering, marketing, and manufacturing. Also see **multidisciplinary team** (Chapter 9).
data dictionary	A dictionary of data contained in a **data flow diagram**. It contains specific information on the data flows and is defined using a formal language (Chapter 6).
data flow diagram	The **intention** of a data flow diagram (DFD) is to model the processing and flow of data inside a system (Chapter 6).
decision matrix	A matrix that is used to evaluate and rank concepts. It integrates both the user needs and the technical merits of different concepts (Chapter 4).
deliverable	Deliverables are entities that are delivered to the project after completion of **tasks.** Also see **activity** (Chapter 10).
derating	A decrease in the maximum amount of power that can be dissipated by a device. The amount of derating is based upon operating conditions, notably increases in temperature (Chapter 8).
descriptive design process	Describes typical activities involved in realizing designs with less emphasis on exact sequencing than a **prescriptive design process** (Chapter 1).
design architecture	The main (Level 1) organization and interconnection of modules in a system (Chapter 5).
design phase	Phase in the **design process** where the technical solution is developed, ultimately producing a detailed system design. Upon its completion, all major systems and subsystems are identified and described using an appropriate model (Chapter 1).
Design process	The steps required to take an idea from concept to realization of the final system. It is a problem-solving methodology that aims to develop a system that best meets the customer's need within given constraints (Chapter 1).
Design space	The space, or collection, of all possible solutions to a design problem (Chapter 2).
detailed design	A phase in the technical design where the problem can be decomposed no further and the identification of elements such as circuit components, logic gates, or software code takes place (Chapter 5).
engineering requirement	A requirement of the system that applies to the technical aspects of the design. An engineering requirement should be abstract, unambiguous, verifiable, traceable, and realistic (Chapter 3).

entity relationship diagram (ERD)	An ERD is used to model database systems. The **intention** of an ERD is to catalog a set of related objects (entities), their attributes, and the relationships between them (Chapter 6).
entity relationship matrix	A matrix that is used to identify relationships between entities in a database system (Chapter 6).
ethics	Philosophy that studies **morality**, the nature of good and bad, and choices to be made (Chapter 11).
event	An event is an occurrence at a specific time and place that needs to be remembered and taken into consideration in the system design (Chapter 6).
event table	A table that is used to store information about **events** in the system. It includes information regarding the event trigger, the source of the event, and process triggered by the event (Chapter 6).
failure function	The failure function $F(t)$ is a mathematical function that provides the probability that a device has failed at time t (Chapter 8).
failure rate	The failure rate $\lambda(t)$ for a device is the expected number of device failures that will occur per unit time (Chapter 8).
fixed costs	Fixed costs are those that are constant regardless of the number of units produced and cannot be directly charged to a process or activity (Chapter 10).
float	The amount of **slippage** that an activity in a project plan can experience without it becoming part of a new **critical path** (Chapter 10).
flowchart	A modeling diagram whose intention is to visually describe a process or algorithm, including its steps and control (Chapter 6).
functional decomposition	A design technique in which a system is designed by determining its overall functionality and then iteratively decomposing it into component subsystems, each with its own functionality (Chapter 5).
functional requirement	A **subsystem design specification** that describes the inputs, outputs, and functionality of a system or component (Chapters 3 and 5).
Gantt chart	Gantt charts are a bar graph representation of a project plan where the activities are shown on a timeline (Chapter 10).
Heisenbugs	Heisenbugs are **bugs** that are not always reproducible with the same input. This is analogous to the Heisenberg uncertainty principle, in which the position of an electron is uncertain (Chapter 7).
high-performance team	A team that significantly outperforms all similar teams. Part of the Katzenbach and Smith team model (Chapter 9).
integration test	An integration test is performed after the units of a system have been constructed and tested. The integration test verifies the operation of the integrated system behavior (Chapter 7).

intention	The intention of a model is the target behavior that it aims to describe (Chapter 6).
interaction view	The interaction view is part of the **Unified Modeling Language**. Its **intention** is to show the interaction between objects. It is characterized by collaboration and sequence diagrams (Chapter 6).
key attribute	An attribute for an entity in a database system that uniquely identifies an instance of the entity (Chapter 6).
lateral thinking	A thought process that attempts to identify creative solutions to a problem. It is not concerned with developing the solution for the problem, or right or wrong solutions. It encourages jumping around between ideas. It is contrasted to **vertical thinking** (Chapter 4).
liable	Required to pay monetary damages according to law (Chapter 11).
maintenance phase	Phase in the **design process** where the system is maintained, upgraded to add new functionality, or design problems are corrected (Chapter 1).
marketing requirement (specifications)	A statement that describes the needs of the customer or end user of a system. It is typically stated in language that the customer would use (Chapters 2 and 3).
matrix test	A matrix test is a test that is suited to cases where the inputs submitted are structurally the same and differ only in their values (Chapter 7).
mean time to failure	The mean time to failure (MTTF) is a mathematical quantity which answers the question, "On average how long does it take for a device to fail?" (Chapter 8).
module	A block, or subsystem, in a design that performs a function (Chapter 5).
morals	The **principles** of right and wrong and the decisions that derive from those principles (Chapter 11).
multidisciplinary team	In general, a multidisciplinary team is one in which the members have complementary skills and the team may have representation from multiple technical disciplines. Also see **cross-functional team** (Chapter 9).
negligence	Failure to exercise caution, which in the case of design could be in not following reasonable standards and rules that apply to the situation (Chapter 11).
network diagram	A network diagram is a directed graph representation of the activities and dependencies between them for a project (Chapter 10).
nominal group technique (NGT)	A formal approach to brainstorming and meeting facilitation. In NGT, each team member silently generates ideas that are reported out in a round-robin fashion so that all members have an opportunity to present their ideas. Concepts are selected by a multivoting scheme, with each member casting a predetermined number of votes for the ideas. The ideas are then ranked and discussed (Chapters 4 and 9).

nondisclosure agreement	An agreement that prevents the signer from disseminating information about a company's products, services, and trade secrets (Chapter 11).
object	Objects represent both data (attributes) and the methods (functions) that can act upon data. An object represents a particular instance of a **class**, which defines the attributes and methods (Chapter 6).
object type	Characteristic of a model used in design. The object type is capable of encapsulating the actual components used to construct the system (Chapter 6).
objective tree	A hierarchical tree representation of the customer's needs. The branches of the tree are organized based upon functional similarity of the needs (Chapter 2).
observability	This principle applies to testing. Observability is the ability to observe any node of a system (Chapter 7).
overspecificity	This refers to applying targets for **engineering requirements** that go beyond what is necessary for the system. Over-specificity limits the size of the **design space** (Chapter 3).
pairwise comparison	A method of systematically comparing all customer needs against each other. A comparison matrix is used for the comparison and the output is a scoring of each of the needs (Appendix B, Chapter 2, and Chapter 4).
parallel system	A system that contains multiple modules performing the same function where a single module would suffice. The overall system functions correctly when any one of the submodules is functioning (Chapter 8).
patent	A patent is a legal device for protecting a design or invention. If a patent is held for a technology, others cannot use it without permission of the owner (Chapter 11).
path-complete coverage	Path-complete coverage is where every possible **processing path** is tested (Chapter 7).
physical view	The physical view is part of the **Unified Modeling Language**. Its **intention** is to demonstrate the physical components of a system and how the logical views map to them. It is characterized by a component and deployment diagram (Chapter 6).
potential team	A team where the sum effort of the team equals that of the individuals working in isolation. Part of the Katzenbach and Smith team model (Chapter 9).
prescriptive design process	An exact process, or systematic recipe, for realizing a system. Prescriptive design processes are often algorithmic in nature and expressed as flowcharts with decision logic (Chapter 1).

principle	Fundamental rules or beliefs that govern behavior, such as the Golden Rule (Chapter 11).
problem identification	The first phase in the design process, where the problem is identified, the customer needs identified, and the project feasibility determined (Chapter 1).
processing path	A processing path is a sequence of consecutive instructions or states encountered while performing a computation. Processing paths are used to develop test cases (Chapter 7).
prototyping and construction phase	Phase in the **design process** in which different elements of the system are constructed and tested. The objective is to model some aspect of the system, demonstrating functionality to be employed in the final realization (Chapter 1).
pseudoteam	An underperforming team for which the sum effort of the team is below that of the individuals working in isolation. Part of the Katzenbach and Smith team model (Chapter 9).
Pugh concept selection	A technique for comparing design concepts to the user needs. It is an iterative process where concepts are scored relative to the needs. Each concept is combined, improved, or removed from consideration in each iteration of the process (Chapter 4).
real team	A team where the sum effort of the team exceeds that of the individuals working in isolation. Part of the Katzenbach and Smith team model (Chapter 9).
redundancy	A design has redundancy if it contains multiple modules performing the same function where a single module would suffice. Redundancy is used to increase **reliability** (Chapter 8).
reliability	Reliability $R(t)$ is the probability that a device is functioning properly (has not failed) at time t (Chapter 8).
requirements specification	A collection of engineering and marketing requirements that a system must satisfy in order for it to meet the needs of the customer or end user. Alternative terms that are used for the requirements specification are the *product design specification* and the *systems requirements specification* (Chapters 1 and 3).
research phase	Phase in the **design process** where research on the basic engineering and scientific principles, related technologies, and existing solutions for the problem are explored (Chapter 1).
reverse engineering	Process where a device or process is taken apart to understand how it works (Chapter 11).

routine design	A formal categorization of design projects. It represents the design of artifacts for which theory and practice are well developed (Chapter 2).
rule-based ethics	Rule-based ethics is based upon a set of rules that can be applied to make decisions. In the strictest form, it is considered to be absolute in terms of governing behavior (Chapter 11).
satisfice	Satisfice means that a solution may meet the design requirements, but not be the optimal solution (Chapter 11).
series system	A system in which the failure of a single component (or subsystem) leads to failure of the overall system (Chapter 8).
situational ethics	Rules that allow decisions to be based on whether they produce the highest good for the person (Chapter 11).
slippage	Refers to an activity in a project plan taking longer than its planned time to complete. See also **critical path** and **float** (Chapter 10).
standards	A standard or established way of doing things. Standards ensure that products work together, from home plumbing fixtures to the modules in a modern computer. They ensure the health and safety of products (Chapter 3).
state	The state of a system represents the net effect of all the previous inputs to the system. Since the state characterizes the history of previous inputs, it is often synonymous with the word *memory* (Chapter 6).
state diagram (machine)	Diagram used to describe systems with memory. It consists of states and transitions between states (Chapter 6).
static view	The static view is part of the **Unified Modeling Language**. The **intention** of the static view is to show the classes in a system and their relationships. The static view is characterized by a class diagram (Chapter 6).
step-by-step test	A step-by-step test case is a prescription for generating a test and checking the results. It is most effective when the test consists of a complex sequence of steps (Chapter 7).
strength and weakness analysis	A technique for the evaluation of potential solutions to a design problem where the strengths and weaknesses are identified (Chapter 4).
strict liability	A form of **liability** that focuses only on the product itself—if the product contains a defect that caused harm, the manufacturer is liable (Chapter 11).
structure charts	Specialized block diagrams for visualizing functional software designs. They employ input, output, transform, coordinate, and composite modules (Chapter 5).
stub	A stub is a device that is used to simulate a subcomponent of a system during testing. Stubs simulate inputs or monitor outputs from the unit under test (Chapter 7).

system integration	Phase in the **design process** where all of the subsystems are brought together to produce a complete working system (Chapter 1).
task	Tasks are actions that accomplish a job as part of a project plan. Also see **activity** and **deliverable** (Chapter 10).
team process guidelines	Guidelines developed by a team that govern their behavior and identify expectations for performance (Chapter 9).
test coverage	Test coverage is the extent to which the test cases cover all possible **processing paths** (Chapter 7).
test phase	Phase in the design process where the system is tested to demonstrate that it meets the requirements (Chapters 1 and 7).
testable	A design is testable when a failure of a component or subsystem can be quickly located. A testable design is easier to debug, manufacture, and service in the field (Chapter 7).
top-down design	An approach to design in which the designer has an overall vision of what the final system must do, and the problem is partitioned into components, or subsystems that work together to achieve the overall goal. Then each subsystem is successively refined and partitioned as necessary. This is contrasted to **bottom-up** design (Chapter 5).
tort	The basis for which a lawsuit is brought forth (Chapter 11).
trade secret	An approach to protecting intellectual property where the information is held secretly, without **patent** protection, so that a competitor cannot access it (Chapter 11).
underspecificity	This refers to a state of the **requirements specification**. When it is underspecified, requirements do not meet the needs of the user and/or embody all of the requirements needed to implement the system (Chapter 3).
Unified Modeling Language (UML)	A modeling language that captures the best practices of object-oriented system design. It encompasses six different system views that can be used to model electrical and computer systems (Chapter 6).
unit test	A unit test is a test of the functionality of a system module in isolation. It establishes that a subsystem performs a single unit of functionality to some specification (Chapter 7).
use-case view	The use-case view is part of the **Unified Modeling Language**. Its **intention** is to capture the overall behavior of the system from the user's point of view and to describe cases in which the system will be used (Chapter 6).
utilitarian ethics	In utilitarian ethics, decisions are based upon the decision that brings about the highest good for all, relative to all other decisions (Chapter 11).

validation	The process of determining whether the requirements meet the needs of the user (Chapter 3).
value	A value is something that a person or group believes to be valuable or worthwhile. Also see **principles** and **morals** (Chapter 11).
variable costs	Variable costs vary depending upon the process or items being produced, and fluctuate directly with the number of units produced (Chapter 10).
variant design	A formal categorization of design projects. It represents the design of existing systems, where the intent is to improve performance or add features (Chapter 2).
verifiable	Refers to a property of an engineering requirement. It means that there should be a way to measure or demonstrate that the requirement is met in the final system realization (Chapter 3).
vertical thinking	A linear, or sequential, thought process that proceeds logically toward the solution of a problem. It seeks to eliminate incorrect solutions. It is contrasted to **lateral thinking** (Chapter 4).
whistleblower	A person who goes outside of the company or organization to report an ethical or safety problem (Chapter 11).
white box test	White box tests are those that are conducted with knowledge of the internal working of the unit under test (Chapter 7).
work breakdown structure	The work breakdown structure (WBS) is a hierarchical breakdown of the tasks and deliverables that need to be completed in order to accomplish a project (Chapter 10).
working group	A group of individuals working in isolation, who come together occasionally to share information. Part of the Katzenbach and Smith team model (Chapter 9).

Appendix B Decision Making with Analytical Hierarchy Process

Making good decisions is a crucial skill at every level. —Peter Drucker

A hallmark of design is that many decisions must be made throughout the process. Typically, there are alternative solutions to a problem from which the best one relative to some criteria is selected. This appendix presents the Analytical Hierarchy Process (AHP) approach to decision making. AHP is a flexible quantitative and qualitative method, applicable to many problems, which provides a numerical score for the alternatives considered. Different aspects of AHP are applied throughout the text, and this appendix is structured to teach AHP via example. A summary of AHP is provided at the conclusion of the appendix.

To apply AHP there must be a decision to be made, criteria against which the decision is based, and a set of competing decisions from which one must be selected. This process is encapsulated in a decision matrix (shown in Table B.1) and therefore is sometimes referred to as the *decision-matrix method*. The row headings are the criteria against which the decision is made and the column headings represent the alternatives. The criteria can have differing levels of importance and their relative weightings are reflected by w_i values in the matrix. The entries in the matrix α_{ij} are ratings for each jth alternative relative to ith criterion. Each alternative receives a score S_j, which is a weighted sum of the ratings computed as

$$S_j = \sum_{i=1}^{m} \omega_i \alpha_{ij}. \tag{1}$$

Table B.1 A decision matrix.

		Alternative 1	Alternative 2	...	Alternative n
Criteria 1	ω_1	α_{11}	α_{12}	...	α_{1n}
Criteria 2	ω_2	α_{21}	α_{22}	...	α_{2n}
$\vdots$	$\vdots$	$\vdots$	$\vdots$	...	$\vdots$
Criteria m	ω_m	α_{m1}	α_{m2}	...	α_{mn}
Score		$S_1 = \sum_{i=1}^{m} \omega_i \alpha_{i1}$	$S_2 = \sum_{i=1}^{m} \omega_i \alpha_{i2}$	...	$S_n = \sum_{i=1}^{m} \omega_i \alpha_{in}$

The steps of AHP are to:
1. Determine the selection criteria.
2. Determine the criteria weightings.
3. Identify and rate alternatives relative to the criteria.
4. Compute scores for the alternatives.
5. Review the decision.

AHP is demonstrated through two examples in which the decision to purchase a car is considered. This example has the benefit of being relatively easy to understand, has readily available public data for supporting the decision, demonstrates the principles of AHP clearly, and is extensible to design problems. The first example demonstrates a straightforward set of criteria, while the second extends the first to include hierarchical criteria.

B.1 Applying AHP for Car Selection

AHP is demonstrated in this section by examining the decision to purchase an automobile.

Step 1: Determine the Selection Criteria

The first step is to brainstorm to identify the criteria against which the decision is made—ideally it is done prior to identification of the alternatives. The pitfall of identifying alternatives first is that the criteria may be selected to bias toward a particular alternative. Assume that the criteria determined are:
- Purchase cost
- Safety
- Design styling
- Brand-name recognition

Step 2: Determine the Criteria Weightings

To determine the weights ω_i, a method known as *pairwise comparison* is applied, where each criterion is systematically compared to all others. For example, the purchase cost is compared to safety, design, and brand name. Likewise, safety is compared to the remaining criteria and so on. A common practice in AHP to apply the following scale for pairwise comparison, and it is used throughout the book for consistency:

1 = equal, 3 = moderate, 5 = strong, 7 = very strong, 9 = extreme.

For example, if one criterion is deemed strongly more important than another, it is assigned a score of 5, while if it is deemed strongly less important, it is assigned the reciprocal value 1/5.

Example comparisons are captured in the comparison matrix in Table B.2. For each cell, the corresponding row criterion is compared to the column criterion. From the first row of the table it is apparent that purchase cost is considered of equal importance to safety, moderately more important than design, and very strongly more important than brand name. By definition, the diagonal elements are assigned values of 1 since each is equally important to itself. The matrix should have the following relationship about the diagonal: $x_{ij} = 1/x_{ji}$.

Table B.2 Pairwise comparison of the selection criteria.

	Purchase cost	Safety	Design	Brand name
Purchase cost	1	1	3	7
Safety	1	1	5	9
Design	1/3	1/5	1	3
Brand name	1/7	1/9	1/3	1

People often make comparisons that are inconsistent. Look at the first row—purchase cost and safety are deemed to be equally important, while cost is moderately more important than design (factor of 3). Yet, in the second row safety is seen as strongly more important than design (by a factor of 5). This is inconsistent, since if we are to believe the first row, then safety would be only moderately more important than design (by a factor of 3)—just as the purchase cost was compared relative to safety in the first row.

An intuitive approach for computing the weight for each criterion is to sum each row. Since a given row represents the comparison of a single criterion to all others, the larger a row sum is the more important it is and the higher the weight it achieves. However, the problem of inconsistency needs to be addressed. There are a number of approaches that can be shown mathematically to reduce the inconsistency in the matrix. A simple method is to take the geometric mean of each row. The geometric mean of a series of numbers, $a_1, \ldots, a_n$, is computed as

$$\text{Geometric mean} = \sqrt[n]{a_1 a_2 \cdots a_n}. \qquad (2)$$

The geometric mean is often used to reduce bias in skewed data. Table B.3 demonstrates how the weights are computed. First, the geometric mean of each row is computed and then the sum of the geometric means is found. The mean values are divided by the sum to produce a normalized set of weights; that is,

$$\sum_i \omega_i = 1.$$

No matter what method is applied to find the weights, they should be normalized to a sum of one.

Table B.3 Weight values computed from the pairwise comparison.

	Purchase cost	Safety	Design	Brand name	Geometric Mean	Weights
Purchase cost	1	1	3	7	2.1	0.37
Safety	1	1	5	9	2.6	0.46
Design	1/3	1/5	1	3	0.7	0.12
Brand name	1/7	1/9	1/3	1	0.3	0.05

The criteria have the following weights $\omega_1 = 0.37$, $\omega_2 = 0.46$, $\omega_3 = 0.12$, and $\omega_4 = 0.05$. These calculations are easily automated with spreadsheet software.

Step 3: Identify and Rate Alternatives Relative to the Criteria

The three competing alternatives to be evaluated are the 2006 model year Honda CR-V, Hyundai Tucson, and Toyota RAV4, which are all small sport-utility vehicles. The ratings of each alternative relative to each criterion, a_{ij}, that make up the body of the decision matrix are determined next. Ideally, quantitative ratings are determined, but in many cases it is necessary to use a more qualitative approach. Higher ratings should reflect a better match to the criteria. Creativity and ingenuity are often needed to determine a proper metric.

Let's examine purchase cost. The vehicle costs are $21,026 (Honda), $18,183 (Hyundai), and $21,989 (Toyota). The purchase cost itself cannot be used directly for a rating metric, since the highest cost would achieve the highest rating, whereas the objective is to minimize cost and reward the lowest cost with the highest rating. An alternative metric is needed. A metric that works when the objective is to minimize a criterion is to compare it to the minimum of all values, using the following ratio

$$\alpha = \frac{\min\{\text{cost}\}}{\text{cost}}. \qquad (3)$$

This assigns a maximum value of 1 to the lowest cost option, which in this case is the Hyundai. The cost ratings are computed to be $\alpha_{11} = 0.86$, $\alpha_{12} = 1$, and $\alpha_{13} = 0.83$. It is important that the ratings relative to each criteria be normalized so that their sum is one. If not, the sum of ratings for each criterion will be different. This would introduce bias by altering the relative weights of the criteria. The normalized ratings are $\alpha_{11} = 0.32$, $\alpha_{12} = 0.37$, and $\alpha_{13} = 0.31$.

Next, a metric for safety is needed. Fortunately, there is real data to draw upon from the U.S. National Highway Transportation Safety Association (*www.safercar.gov*) which rates vehicles in multiple categories on a 5-point scale. The average rating for each car is $\alpha_{21} = 4.8$ (Honda), $\alpha_{22} = 4.8$ (Hyundai), and $\alpha_{23} = 4.6$ (Toyota), producing the following normalized values $\alpha_{21} = 0.34$, $\alpha_{22} = 0.34$, and $\alpha_{23} = 0.32$.

The rating of the cars relative to the design styling criterion is considered next. This requires a more subjective approach than the previous two criteria. To quantify the subjective evaluation pairwise comparison is again applied to determine the relative value of one automobile's design to another. Pairwise comparison for the design styling criteria is shown in Table B.4. The 2006 CR-V has older styling, resulting in a much lower design styling rating than the others.

Table B.4 Pairwise comparison of design styling to determine ratings.

	Honda CRV	Hyundai Tucson	Toyota RAV4	Design Rating
Honda CRV	1	1/3	1/5	0.11
Hyundai Tucson	3	1	1/2	0.31
Toyota RAV4	5	2	1	0.58

Finally, the ratings for brand-name recognition are determined by pairwise comparison as shown in Table B.5.

Table B.5 Pairwise comparison of brand name to determine ratings.

	Honda CRV	Hyundai Tucson	Toyota RAV4	Brand name Rating
Honda CRV	1	4	1	0.44
Hyundai Tucson	1/4	1	1/4	0.12
Toyota RAV4	1	4	1	0.44

Note that the pairwise comparison method can be somewhat time-consuming since many comparisons must be made. For making subjective estimates, a faster approach is to use scoring rubric that reflects how well each of the alternatives meet the criterion; for example,

1 = does not meet criterion, 5 = partially meets criterion, 9 = completely meets criterion.

Keep in mind that pairwise comparison has the advantage of a systematic comparison of alternatives, and this reduces inconsistency in ratings.

Step 4: Compute Scores for the Alternatives

The decision matrix is built and the overall weighted scores for the alternatives are computed as shown in Table B.6.

Table B.6 The decision matrix.

		Honda CR-V	Hyundai Tucson	Toyota RAV4
Cost	0.37	0.32	0.37	0.31
Safety	0.46	0.34	0.34	0.32
Design styling	0.12	0.11	0.31	0.58
Brand name	0.05	0.44	0.12	0.44
Score		0.31	0.34	0.35

Step 5: Review the Decision

The result is a set of numerical scores for the alternatives, and if all work is done properly the final scores should sum to one. In this case there is not much difference between the scores, and a simple decision based upon the maximum value would lead to selection of the RAV4. Since there is not much of a difference, all three are good choices according to the selection criteria. The decision matrix allows the examination of different scenarios, as the weights and ratings can be varied to see how they affect the overall scores. The RAV4 beats the Tucson primarily because of styling and brand name, which are both subjective ratings. This might warrant revisiting those ratings to see how they affect the decision. The RAV4 barely beats out the Tucson, so if cost and safety are truly more important, the Tucson is a better decision, as it beats the RAV4 in both of those categories.

B.2 Hierarchical Decision Criteria

The example in the previous section had fairly simple selection criteria and it arguably missed some important ones, such as the operating costs. Further, when considering operating cost it becomes apparent that it has multiple dimensions, or subcriteria, such as fuel and insurance costs. Similarly, the design styling criterion can be subdivided into interior and exterior design. Thus there is a hierarchy in the criteria, giving rise to the name analytical hierarchy process. The following extends the previous example to include the use of hierarchical criteria.

Step 1: Determine the Selection Criteria

Assume that the criteria have been expanded to be more realistic as shown below:
- Purchase cost
- Operating costs
    - Fuel; miles per gallon (MPG)
    - Insurance
- Safety
- Design styling
    - Interior
    - Exterior
- Brand-name recognition

Step 2: Determine the Criteria Weightings

Pairwise comparison is again used to determine the weights. The difference now is that this comparison is done at each level in the hierarchy. First, purchase cost, operating costs, safety, design, and brand-name recognition are compared. Then fuel and insurance costs are compared to each other since they are at the same level and grouping in the hierarchy, as are exterior and interior design. Assume that the pairwise comparisons produce the following criteria weights:

- Purchase cost (0.33)
- Operating costs (0.11)
    - MPG (0.67)
    - Insurance (0.33)
- Safety (0.40)
- Design Styling (0.12)
    - Interior (0.50)
    - Exterior (0.50)
- Brand-name recognition (0.04)

Step 3: Identify and Rate Alternatives Relative to the Criteria

To utilize the results from the previous section, it is only necessary to compute scores relative to the new criteria. Using mileage per gallon as a metric for ratings fuel costs is straightforward, since the two are directly proportional. This data is readily available and is 24, 23, and 25 miles per gallon for the CR-V, Tucson, and RAV4 respectively. This produces normalized ratings of 0.33, 0.32, and 0.35 respectively.

In terms of insurance costs, the National Highway Transportation Safety Institute publishes the relative average loss per insured vehicle, a rating that is used by insurance companies to determine insurance costs. A score of 100 is the industry average—those exceeding 100 are above the average and those less than 100 are below it. The publicly available ratings are 93, 100, and 112 for the CR-V, Tucson, and RAV4 respectively. Again, the objective is to minimize these values so the metric in equation (3) is used. This produces normalized ratings of 0.36, 0.34, and 0.30 respectively.

Styling is subjective and ratings could be determined using a scoring rubric or the pairwise comparison method. Assume that this produces the scores shown in Table B.7.

Step 4: Compute Scores for the Alternatives

The decision matrix shown in Table B.7 reflects the hierarchical criteria. The scores are computed slightly differently when subcriteria are involved. For example, the score for operating costs of the Honda is based upon the two subcriteria as (0.67 × 0.33 + 0.33 × 0.36) = 0.34. The value of 0.34 is then the rating reflected in the table for the operating costs of this vehicle. Then to get the weighted score for operating costs, this result of 0.33 is multiplied by the weight of 0.11, exactly as in the previous example.

Table B.7 The decision matrix.

		Honda CR-V	Hyundai Tucson	Toyota RAV4
Purchase Cost	0.33	0.32	0.37	0.31
Operating Cost	0.11	0.34	0.32	0.33
Fuel	0.67	0.33	0.32	0.35
Insurance	0.33	0.36	0.34	0.30
Safety	0.40	0.34	0.34	0.32
Design Styling	0.12	0.20	0.40	0.40
Interior	0.60	0.20	0.40	0.40
Exterior	0.40	0.20	0.40	0.40
Brand-name	0.04	0.44	0.12	0.44
Score		0.32	0.35	0.33

Step 5: Review the Decision

With this expanded set of criteria, there is a shift in the overall ranking and the Hyundai Tucson edges out the Toyota RAV4. Frankly, all three are good choices and reflect the highly competitive car market. A comparison of vehicles across classes, such as sport-utility, mid-size, and luxury cars would likely result in a much larger differential in the scoring on the basis of the weighted criteria used here.

B.3 Summary and Further Reading

The analytical hierarchy process is an effective tool for comparing competing alternatives that integrates both quantitative and qualitative judgments. In order to compare alternatives, criteria for comparison are selected and the relative weights of the criteria assigned. In terms of selecting the criteria and their weightings, remember the following:

- Ideally, the criteria are selected prior to identification of the alternatives. This avoids the trap of selecting the criteria to support a preconceived decision.
- There is no single correct way to compute the criterion weights. However, the pairwise comparison approach is well accepted. It can be a bit time-consuming, particularly when a large number of criteria are to be compared. If used properly, pairwise comparison reduces bias in the decision.
- There are alternatives to pairwise comparison. One is to rank the criterion from most important to least important and then assign a relative weighting based upon the ranking. Another alternative is to utilize a scoring rubric using a semiqualitative scale, as 1 = low importance, 5 = medium importance, and 9 = extremely important.
- Regardless of how the weights are selected, the sum of the weights should be normalized to a value of one.

Once the criteria and weights are found, ratings for each option relative to the criteria are determined. When computing the ratings, remember the following:
- If quantitative data is available, it should be used. It is admittedly hard to determine for many problems.
- Appropriate metrics need to be used for rating quantitative data.
- If quantitative data is unavailable, qualitative judgments must be made and quantified. Again, the pairwise comparison is a good tool for quantifying such judgments. Alternatively, a rubric for rating the alternatives against the criteria can be used.
- The ratings relative to each criterion need to be normalized so that their sum is equal to one.

The final result is a decision matrix and scores for each of the alternatives. This matrix shows the quantified criteria, allowing a rational approach to decision making and minimizing emotional factors in the decision.

AHP was originally developed by Thomas Satay [Sat88] and has found wide acceptance since. There are software packages available, such as Expert Choice, that allow for rapid comparison of alternatives and evaluation.

Appendix C Component Failure Rate Data

This appendix contains failure rate data for selected electronic components in support of Chapter 8. The information was abstracted from the Military Handbook for Reliability Prediction [MIL-HDBK-217F] and contains information for devices that are relevant to this book. The information is a close facsimile to MIL-HDBK-217F, but in some instances comments are added to help interpret the information. Furthermore, MIL-HDBK-217F tends to present both empirical estimation formulas and tabular data computed from the formulas. The tabular data is generally not included here for simplicity of presentation, and is readily obtained from the formulas supplied. Data is presented for the following devices:

- Analog components: resistors and capacitors.
- Discrete semiconductors: diodes, bipolar transistors, and field-effect transistors.
- Microcircuits: gate/logic arrays and microprocessors.

C.1 Environmental Use

All of the devices presented in this appendix have an environmental factor that is used to estimate the failure rate. The environmental factors are based on the 14 categories in Table C.1.

Table C.1 Environmental symbols and descriptions taken directly from MIL-HDBK-217F.

Environment	Description
G_B – Ground, Benign	Nonmobile, temperature- and humidity-controlled environments readily accessible to maintenance; includes laboratory instruments and test equipment, medical electronic equipment, business and scientific computer complexes, and missile and support equipment in ground silos.
G_F – Ground, Fixed	Moderately controlled environments such as installation in permanent racks with adequate cooling air and possible installation in unheated buildings; includes permanent installation of air traffic control radar and communication facilities.
G_M – Ground, Mobile	Equipment installed in wheeled or tracked vehicles and equipment manually transported; includes tactical missile ground support equipment, mobile communication equipment, tactical fire direction systems, handheld communications equipment, laser designators, and range finders.

Table C.1 Environmental symbols and descriptions taken directly from MIL-HDBK-217F, cont'd.

Environment	Description
N_S – Naval, Sheltered	Includes sheltered or below-deck conditions on surface ships and equipment installed in submarines.
N_U – Naval, Unsheltered	Unprotected surface shipborne equipment exposed to weather conditions and equipment immersed in salt water. Includes sonar equipment and equipment installed on hydrofoil vessels.
A_{IC} – Airborne, Inhabited Cargo	Typical conditions in cargo compartments that can be occupied by an air crew. Environmental extremes of pressure, temperature, shock, and vibration are minimal. Examples include long-mission aircraft such as the C130, C5, B52, and C141. This category also applies to inhabited areas in lower-performance smaller aircraft such as the T38.
A_{IF} – Airborne, Inhabited Fighter	Same as A_{IC} but installed on high-performance aircraft such as fighters and interceptors. Examples include the F15, F16, F111, F/A18, and A10 aircraft.
A_{UC} – Airborne, Uninhabited Cargo	Environmentally uncontrolled areas that cannot be inhabited by crew during flight. Environmental extremes of pressure, temperature, and shock may be severe. Examples include uninhabited areas of long-mission aircraft such as the C130, C5, B52, and C141. This category also applies to uninhabited areas in lower-performance smaller aircraft such as the T38.
A_{UF} - Airborne, Uninhabited Fighter	Same as A_{UC} but installed on high-performance aircraft such as fighters and interceptors. Examples include the F15, F16, F111, and A10 aircraft.
A_{RW} – Airborne, Rotary Winged	Equipment installed on helicopters. Applies to both internally and externally mounted equipment such as laser designators, fire control systems, and communications equipment.
S_F – Space, Flight	Earth orbital. Approaches benign ground conditions. Vehicle neither under powered flight nor in atmospheric reentry; includes satellites and shuttles.
M_F – Missile, Flight	Conditions related to powered flight of air breathing missiles, cruise missiles, and missiles in unpowered free flight.
M_L – Missile, Launch	Severe conditions related to missile launch (air, ground, and sea), space vehicle boost into orbit, and vehicle re-entry and landing by parachute. Also applies to solid rocket motor propulsion powered flight, and torpedo and missile launch from submarines.
C_L – Cannon, Launch	Extremely severe conditions related to cannon launching of 155-mm and 5-inch guided projectiles. Conditions apply to the projectile from launch to target impact.

C.2 Analog Components: Resistors and Capacitors

This section contains failure rate data for resistors and capacitors.

C.2.1 Resistors: Fixed Composition, Fixed Film, and Wirewound

The failure rate is given by the following relationship:

$$\lambda = \lambda_b \pi_R \pi_Q \pi_E \, \frac{\text{failures}}{10^6 \text{ hours}}.$$

λ_b — Base Failure Rate

Resistor Type	λ_b $\left(T = \text{ambient temperature in °C}, S = \dfrac{\text{operating power}}{\text{device power rating}}\right)$
Fixed Composition	$\lambda_b = 4.5 \times 10^{-9} \exp\left[12\left(\dfrac{T+273}{343}\right)\right] \exp\left[\dfrac{S}{0.6}\left(\dfrac{T+273}{273}\right)\right]$
Fixed Film	$\lambda_b = 3.25 \times 10^{-4} \exp\left[\left(\dfrac{T+273}{343}\right)^3\right] \exp\left[S\left(\dfrac{T+273}{273}\right)\right]$
Wirewound	$\lambda_b = 0.0031 \exp\left[\left(\dfrac{T+273}{398}\right)^{10}\right] \exp\left\{\left[S\left(\dfrac{T+273}{273}\right)\right]^{1.5}\right\}$

Resistance Factor — π_R

Fixed Composition or Fixed Film Resistors		Wirewound Resistors	
Resistance Range	π_R	Resistance Range	π_R
$\leq 100 \text{ k}\Omega$	1.0	$\leq 10 \text{ k}\Omega$	1.0
$100 \text{ k}\Omega$ to $\leq 1 \text{ M}\Omega$	1.1	$10 \text{ k}\Omega$ to $\leq 100 \text{ k}\Omega$	1.7
$1 \text{ M}\Omega$ to $\leq 10 \text{ M}\Omega$	1.6	$100 \text{ k}\Omega$ to $\leq 1 \text{ M}\Omega$	3.0
$> 10 \text{ M}\Omega$	2.5	$> 1 \text{ M}\Omega$	5.0

Quality Factor—π_Q

Quality	π_Q
S	0.03
R	0.1
P	0.3
M	1.0
MIL-R	5.0
Lower	15

Environmental factor—π_E

Environment	π_E—Fixed Composition	π_E—Fixed Film	π_E—Wirewound
G_B	1	1	1
G_F	3	2	2
G_M	8	8	11
N_S	5	4	5
N_U	13	14	18
A_{IC}	4	4	15
A_{IF}	5	8	18
A_{UC}	7	10	28
A_{UF}	11	18	35
A_{RW}	19	19	27
S_F	0.50	0.20	0.80
M_F	11	10	14
M_L	27	28	38
C_L	490	510	610

C.2.2 Capacitors: Fixed, Ceramic, and General Purpose

The failure rate is given by the following relationship:

$$\lambda = \lambda_b \pi_{CV} \pi_Q \pi_E \frac{\text{failures}}{10^6 \text{ hours}}.$$

For all factors T = ambient temperature (°C) and $S = \frac{\text{operating voltage}}{\text{rated voltage}}$, where the operating voltage is the sum of the DC and peak AC voltages.

Base Failure Rate – λ_b

$$\lambda_b = 0.0003 \left[\left(\frac{S}{0.3} \right)^3 + 1 \right] \exp\left(\frac{T+273}{358} \right), \quad \text{for } T = 85°C \text{ max rated}$$

$$\lambda_b = 0.0003 \left[\left(\frac{S}{0.3} \right)^3 + 1 \right] \exp\left(\frac{T+273}{398} \right), \quad \text{for } T = 125°C \text{ max rated}$$

$$\lambda_b = 0.0003 \left[\left(\frac{S}{0.3} \right)^3 + 1 \right] \exp\left(\frac{T+273}{423} \right), \quad \text{for } T = 150°C \text{ max rated}$$

Capacitance Factor – π_{CV}

$\pi_{CV} = 0.41 \times C^{0.11}$

C = capacitance

Quality Factor – π_Q

Quality	π_Q
S	0.030
R	0.10
P	0.30
M	1.0
L	3.0
MIL	3.0
Lower	10

Environmental Factor – π_E

Environment	π_E
G_B	1
G_F	2
G_M	9
N_S	5
N_U	15
A_{IC}	4
A_{IF}	4
A_{UC}	8
A_{UF}	12
A_{RW}	20
S_F	0.40
M_F	13
M_L	34
C_L	610

C.3 Microelectronic Devices

For all microelectronic devices it is necessary to compute the junction temperature (T_J) of the silicon in order to determine the temperature factor. The junction temperature is determined as follows:

$$T_J = T_A + \theta_{JA} P_D$$

T_A = ambient temperature

θ_{JA} = junction to ambient thermal resistance (obtained from manufacturer data sheet)

P_D = power dissipated in the device

Authors' Note: The equations above are slightly different than those found in MIL-HDBK-217F. The junction-to-ambient thermal resistance is used here, instead of junction to case as in the original. In addition, the ambient temperature is used in place of the case temperature. This is more general and is consistent with the presentation in Chapter 8. The junction-to-case resistance could be used, along with the case temperature. See Section 8.2 for more detailed coverage of this thermal model.

The part quality descriptors in Table C.2 are used to find the quality factors.

Table C.2 Part quality descriptors for microelectronic devices.

JANTXV	Full device testing as specified by the MIL-S-19500 specification, including screening and Groups A, B, and C.
JANTX	Identical to JANTXV, except does not include the 100% precap visual inspection contained in screening.
JAN	Testing as defined by MIL-S-19500, including Groups A, B, and C, but not including screening.
Lower	All hermetically packaged devices.
Plastic	All devices encapsulated with organic materials.

C.3.1 Diodes: Low Frequency

The failure rate is given by the following relationship:

$$\lambda = \lambda_b \pi_R \pi_Q \pi_E \frac{\text{failures}}{10^6 \text{ hours}}.$$

Base Failure Rate — λ_b

Diode Type/Application	λ_b
General-purpose analog	0.0038
Switching	0.0010
Power rectifier, fast recovery	0.069
Power rectifier, Schottky power diode	0.0030
Power rectifier with high-voltage stacks	0.0050/junction
Transient suppressor/varistor	0.0013
Current regulator	0.0034
Voltage regulator and voltage reference (avalanche and zener)	0.0020

Temperature Factor — π_T

For general-purpose analog, switching, fast recovery, power rectifier, and transient suppressor applications

$$\pi_T = \exp\left[-3091\left(\frac{1}{T_J + 273} - \frac{1}{298}\right)\right].$$

For voltage regulator, voltage reference, and current regulator applications

$$\pi_T = \exp\left[-1925\left(\frac{1}{T_J + 273} - \frac{1}{298}\right)\right].$$

Electrical Stress Factor—π_S

$\pi_{s'} = 0.054, \quad V_S \leq 0.3$

$\pi_{s'} = V_S^{2.43}, \quad 0.3 < V_S \leq 1$

V_s = voltage stress ratio = $\dfrac{\text{applied voltage}}{\text{rated voltage}}$

Applied voltage is the diode reverse voltage

Quality Factor—π_Q

Quality	π_Q
JANTXV	0.7
JANTX	1.0
JAN	2.4
Lower	5.5
Plastic	8.0

Contact Construction Factor—π_C

Contact Construction	π_C
Metallurgically bonded	1.0
Nonmetallurgically bonded and spring-loaded contacts	2.0

Environmental Factor—π_E

Environment	π_E
G_B	1
G_F	6
G_M	9
N_S	9
N_U	19
A_{IC}	13
A_{IF}	29
A_{UC}	20
A_{UF}	43
A_{RW}	24
S_F	0.50
M_F	14
M_L	32
C_L	320

C.3.2 Diodes: High Frequency (Microwave, RF)

The failure rate is given by the following relationship:

$$\lambda = \lambda_b \pi_T \pi_A \pi_R \pi_Q \pi_E \frac{\text{failures}}{10^6 \text{ hours}}.$$

Base Failure Rate — λ_b

Diode Type	λ_b
Si IMPATT ($\leq$ 35 GHz)	0.22
Gunn/Bulk Effect	0.18
Tunnel and Back	0.0023
PIN	0.0081
Schottky Barrier	0.027
Varactor and Step Recovery	0.0025

Application Factor — π_A

Application	π_A
Varactor, voltage control	0.50
Varactor, multiplier	2.5
All other diodes	1.0

Quality Factor — π_Q

Quality	π_Q Not Shottky	π_Q Shottky
JANTXV	0.5	0.5
JANTX	1.0	1.0
JAN	5.0	1.8
Lower	25.0	2.5
Plastic	50.0	-

Temperature Factor — π_T

All types except IMPATT

$$\pi_T = \exp\left[-2100\left(\frac{1}{T_J + 273} - \frac{1}{298}\right)\right],$$

and for IMPATT

$$\pi_T = \exp\left[-5260\left(\frac{1}{T_J + 273} - \frac{1}{298}\right)\right].$$

Power Rating Factor — π_R

$\pi_R = 0.326 \ln(P_R) - 0.25$, PIN diodes
$\pi_R = 1.0$, all other diodes

Environmental Factor — π_E

Environment	π_E
G_B	1
G_F	2
G_M	5
N_S	4
N_U	11
A_{IC}	4
A_{IF}	5
A_{UC}	7
A_{UF}	12
A_{RW}	16
S_F	0.5
M_F	9
M_L	24
C_L	250

C.3.3 Transistors: Bipolar Junction, Low Frequency (≤ 200 MHz)

The failure rate is given by the following relationship:

$$\lambda = \lambda_b \pi_T \pi_A \pi_R \pi_S \pi_Q \pi_E \frac{\text{failures}}{10^6 \text{ hours}}.$$

Base Failure Rate — λ_b

Type	λ_b
NPN or PNP	0.00074

Temperature Factor — π_T

$$\pi_T = \exp\left[-2114\left(\frac{1}{T_J+273} - \frac{1}{298}\right)\right]$$

Application Factor — π_A

Application	π_A
Linear Amplification	1.50
Switching	0.70

Power Rating Factor — π_R

$\pi_R = 0.43$, if $P_R \leq 0.1\text{W}$

$\pi_R = P_R^{0.37}$, if $P_R > 0.1\text{W}$

where P_R = rated power

Voltage Stress Factor — π_S

$\pi_S = 0.045 \exp(3.1 V_s), \quad 0 \leq V_s \leq 1$

$V_s = \dfrac{\text{applied } V_{CE}}{\text{rated } V_{CEO}}$

V_{CE} = voltage, collector to emitter

V_{CEO} = voltage, collector to emitter, base open

Part Quality Factor — π_Q

Quality	π_Q
JANTXV	0.7
JANTX	1.0
JAN	2.4
Lower	5.5
Plastic	8.0

Environmental Factor — π_E

Environment	π_E
G_B	1
G_F	6
G_M	9
N_S	9
N_U	19
A_{IC}	13
A_{IF}	29
A_{UC}	20
A_{UF}	43
A_{RW}	24
S_F	0.50
M_F	14
M_L	32
C_L	320

C.3.4 Transistors: Bipolar Junction, High Frequency (> 200 MHz), Low Noise (Power ≤ 1W)

The failure rate is given by the following relationship:

$$\lambda = \lambda_b \pi_T \pi_R \pi_S \pi_Q \pi_E \frac{\text{failures}}{10^6 \text{ hours}}.$$

Base Failure Rate—λ_b

Type	λ_b
All types	0.18

Power Rating Factor—π_R

$\pi_R = 0.43$, if $P_R \leq 0.1\text{W}$

$\pi_R = P_R^{0.37}$, if $P_R > 0.1\text{W}$

where P_R = rated power

Part Quality Factor—π_Q

Quality	π_Q
JANTXV	0.5
JANTX	1.0
JAN	2.0
Lower	5.0

Temperature Factor—π_T

$$\pi_T = \exp\left[-2114\left(\frac{1}{T_J+273} - \frac{1}{298}\right)\right]$$

Voltage Stress Factor—π_S

$\pi_s = 0.045 \exp(3.1 V_s)$, $0 \leq V_s \leq 1$

$V_s = \dfrac{\text{applied } V_{CE}}{\text{rated } V_{CEO}}$

V_{CE} = voltage, collector to emitter

V_{CEO} = voltage, collector to emitter, base open

Environmental Factor—π_E

Environment	π_E
G_B	1
G_F	2
G_M	5
N_S	4
N_U	11
A_{IC}	4
A_{IF}	5
A_{UC}	7
A_{UF}	12
A_{RW}	16
S_F	0.50
M_F	9
M_L	24
C_L	250

C.3.5 Transistors: Field Effect, Low Frequency (≤ 400MHz)

The failure rate is given by the following relationship:

$$\lambda = \lambda_b \pi_T \pi_A \pi_Q \pi_E \; \frac{\text{failures}}{10^6 \text{ hours}}.$$

Base Failure Rate — λ_b

Type	λ_b
MOSFET	0.012
JFET	0.0045

Temperature Factor — π_T

$$\pi_T = \exp\left[-1925\left(\frac{1}{T_J+273} - \frac{1}{298}\right)\right]$$

Application Factor — π_A

Application (P_R = rated output power)	π_A
Linear Amplification ($P_R < 2\,W$)	1.50
Small Signal Switching	0.70
Power FETs (Nonlinear, ($P_R < 2\,W$)	
$2 \leq P_R < 5\,W$	2.0
$5 \leq P_R < 50\,W$	4.0
$50 \leq P_R < 250\,W$	8.0
$P_R \geq 250\,W$	10.0

Part Quality Factor — π_Q

Quality	π_Q
JANTXV	0.7
JANTX	1.0
JAN	2.4
Lower	5.5
Plastic	8.0

Environmental Factor — π_E

Environment	π_E
G_B	1
G_F	6
G_M	9
N_S	9
N_U	19
A_{IC}	13
A_{IF}	29
A_{UC}	20
A_{UF}	43
A_{RW}	24
S_F	0.50
M_F	14
M_L	32
C_L	320

C.3.6 Transistors: Field Effect, High Frequency (> 400MHz), Low Power (≤ 300mW)

The failure rate is given by the following relationship:

$$\lambda = \lambda_b \pi_T \pi_Q \pi_E \frac{\text{failures}}{10^6 \text{ hours}}.$$

Base Failure Rate—λ_b

Type	λ_b
MOSFET	0.060
JFET	0.023

Part Quality Factor—π_Q

Quality	π_Q
JANTXV	0.5
JANTX	1.0
JAN	2.0
Lower	5.0

Temperature Factor—π_T

$$\pi_T = \exp\left[-1925\left(\frac{1}{T_J + 273} - \frac{1}{298}\right)\right]$$

Environmental Factor—π_E

Environment	π_E
G_B	1
G_F	2
G_M	5
N_S	4
N_U	11
A_{IC}	4
A_{IF}	5
A_{UC}	7
A_{UF}	12
A_{RW}	16
S_F	0.50
M_F	9
M_L	24
C_L	250

C.3.7 Microcircuits: Gate/Logic Arrays and Microprocessors

Includes the following devices:
1. Bipolar Devices, Digital and Linear Gate/Logic Arrays
2. MOS Devices, Digital and Linear Gate/Logic Arrays
3. Field Programmable Logic Array (PLA) and Programmable Array Logic (PAL)
4. Microprocessors

The failure rate is given by the following relationship:

$$\lambda = (C_1 \pi_T + C_2 \pi_E) \pi_Q \pi_L \frac{\text{failures}}{10^6 \text{ hours}}.$$

C_1 = Complexity Failure Rate for Bipolar Devices (Digital and Linear Gate/Logic)

Digital		Linear		PLA/PAL	
No. Gates	C_1	No. Transistors	C_1	No. Gates	C_1
1 to 100	0.0025	1 to 100	0.010	Up to 200	0.010
101 to 1,000	0.0050	101 to 300	0.020	201 to 1,000	0.021
1,001 to 3,000	0.010	301 to 1,000	0.040	1,001 to 5,000	0.042
3,001 to 10,000	0.020	1,001 to 10,000	0.060		
10,001 to 30,000	0.040				
30,001 to 60,000	0.080				

C_1 = Complexity Failure Rate for MOS Devices (Digital and Linear Gate/Logic)

Digital		Linear		MOS	
No. Gates	C_1	No. Transistors	C_1	No. Gates	C_1
1 to 100	0.010	1 to 100	0.010	Up to 500	0.00085
101 to 1,000	0.020	101 to 300	0.020	501 to 1,000	0.0017
1,001 to 3,000	0.040	301 to 1,000	0.040	1,001 to 5,000	0.0034
3,001 to 10,000	0.080	1,001 to 10,000	0.060	5,001 to 20,000	0.0068
10,001 to 30,000	0.16				
30,001 to 60,000	0.29				

Appendix C Component Failure Rate Data

C_1 = Complexity Failure Rate for Microprocessors

Number of Bits	C_1 – Bipolar	C_1 – MOS
Up to 8	0.060	0.14
Up to 16	0.12	0.28
Up to 32	0.24	0.56

C_2 = Package Failure Rate for all Microcircuits. N_p = number of pins on package.

$C_2 = 2.8 \times 10^{-4}(N_p)^{1.08}$, Hermetic: DIPs with solder or weld seal, SMT
$C_2 = 9.0 \times 10^{-5}(N_p)^{1.51}$, DIPs with glass seal
$C_2 = 3.0 \times 10^{-5}(N_p)^{1.82}$, Flatpacks with axial leads on 50 mil centers
$C_2 = 2.8 \times 10^{-4}(N_p)^{2.01}$, Cans
$C_2 = 3.0 \times 10^{-5}(N_p)^{1.08}$, Nonhermetic: DIPs, PGA, SMT (leaded and nonleaded)

Temperature Factor — π_T

$$\pi_T = 0.1 \times \exp\left[-\frac{E_a}{8.617 \times 10^{-5}}\left(\frac{1}{T_J + 273} - \frac{1}{296}\right)\right]$$

The activation energies (E_a) for different technologies are given in the table below.

Technology	TTL, ASTLL, CML, HTTL, FTLL, DTL, ECL, ALSTTL	F, LTTL, STTL	BiCMOS LSTTL	Digital MOS, VHSIC CMOS	Linear (Bipolar and MOS)	Memories (Bipolar and MOS), NMOS
E_a	0.4	0.45	0.5	0.35	0.65	0.6

Design for Electrical and Computer Engineers

Learning Factor—π_L

Years in Production	π_L
≤ .1	2.0
0.5	1.8
1.0	1.5
1.5	1.2
≥ 2.0	1.0

Part Quality Factor—π_Q

Quality	π_Q
S	0.25
B	1.0
B-1	2.0

Environmental Factor—π_E

Environment	π_E
G_B	0.50
G_F	2
G_M	4
N_S	4
N_U	6
A_{IC}	4
A_{IF}	5
A_{UC}	5
A_{UF}	8
A_{RW}	8
S_F	0.50
M_F	5
M_L	12
C_L	220

Appendix D Manufacturer Data Sheets

This appendix contains manufacturer data sheets for selected electronic components in support of Chapter 8. The following data sheets are included:

- 1N4001 Rectifier Diode.
- 2N3904 General-Purpose Amplifier NPN Transistor.
- CD4001BM/BC Quad 2-Input NOR Gate.
- LM741 General Purpose Operational Amplifier.

1N4001 - 1N4007

Features
- Low forward voltage drop.
- High surge current capability.

DO-41
COLOR BAND DENOTES CATHODE

General Purpose Rectifiers

Absolute Maximum Ratings* T_A = 25°C unless otherwise noted

Symbol	Parameter	Value							Units
		4001	4002	4003	4004	4005	4006	4007	
V_{RRM}	Peak Repetitive Reverse Voltage	50	100	200	400	600	800	1000	V
$I_{F(AV)}$	Average Rectified Forward Current, .375" lead length @ T_A = 75°C	1.0							A
I_{FSM}	Non-repetitive Peak Forward Surge Current 8.3 ms Single Half-Sine-Wave	30							A
T_{stg}	Storage Temperature Range	-55 to +175							°C
T_J	Operating Junction Temperature	-55 to +175							°C

*These ratings are limiting values above which the serviceability of any semiconductor device may be impaired.

Thermal Characteristics

Symbol	Parameter	Value	Units
P_D	Power Dissipation	3.0	W
$R_{\theta JA}$	Thermal Resistance, Junction to Ambient	50	°C/W

Electrical Characteristics T_A = 25°C unless otherwise noted

Symbol	Parameter	Device							Units
		4001	4002	4003	4004	4005	4006	4007	
V_F	Forward Voltage @ 1.0 A	1.1							V
I_{rr}	Maximum Full Load Reverse Current, Full Cycle T_A = 75°C	30							µA
I_R	Reverse Current @ rated V_R T_A = 25°C T_A = 100°C	5.0 500							µA µA
C_T	Total Capacitance V_R = 4.0 V, f = 1.0 MHz	15							pF

©2003 Fairchild Semiconductor Corporation   1N4001-1N4007, Rev. C1

National Semiconductor

Discrete Power & Signal Technologies

2N3904

TO-92

MMBT3904
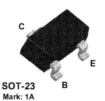
SOT-23
Mark: 1A

MMPQ3904

SOIC-16

PZT3904
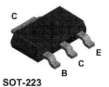
SOT-223

NPN General Purpose Amplifier

This device is designed as a general purpose amplifier and switch. The useful dynamic range extends to 100 mA as a switch and to 100 MHz as an amplifier. Sourced from Process 23.

Absolute Maximum Ratings*   TA = 25°C unless otherwise noted

Symbol	Parameter	Value	Units
V_{CEO}	Collector-Emitter Voltage	40	V
V_{CBO}	Collector-Base Voltage	60	V
V_{EBO}	Emitter-Base Voltage	6.0	V
I_C	Collector Current - Continuous	200	mA
T_J, T_{stg}	Operating and Storage Junction Temperature Range	-55 to +150	°C

*These ratings are limiting values above which the serviceability of any semiconductor device may be impaired.

NOTES:
1) These ratings are based on a maximum junction temperature of 150 degrees C.
2) These are steady state limits. The factory should be consulted on applications involving pulsed or low duty cycle operations.

NPN General Purpose Amplifier
(continued)

Thermal Characteristics TA = 25°C unless otherwise noted

Symbol	Characteristic	Max		Units
		2N3904	*PZT3904	
P_D	Total Device Dissipation Derate above 25°C	625 5.0	1,000 8.0	mW mW/°C
$R_{\theta JC}$	Thermal Resistance, Junction to Case	83.3		°C/W
$R_{\theta JA}$	Thermal Resistance, Junction to Ambient	200	125	°C/W

Symbol	Characteristic	Max		Units
		**MMBT3904	MMPQ3904	
P_D	Total Device Dissipation Derate above 25°C	350 2.8	1,000 8.0	mW mW/°C
$R_{\theta JA}$	Thermal Resistance, Junction to Ambient Effective 4 Die Each Die	357	 125 240	°C/W °C/W °C/W

*Device mounted on FR-4 PCB 36 mm X 18 mm X 1.5 mm; mounting pad for the collector lead min. 6 cm².

**Device mounted on FR-4 PCB 1.6" X 1.6" X 0.06."

Typical Characteristics

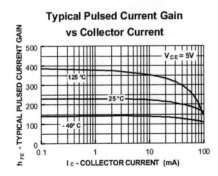

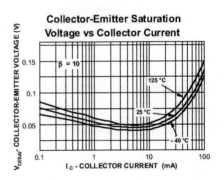

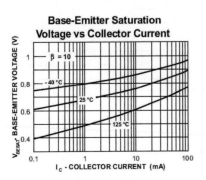

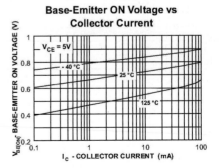

CD4001BM/CD4001BC Quad 2-Input NOR Buffered B Series Gate
CD4011BM/CD4011BC Quad 2-Input NAND Buffered B Series Gate

March 1988

General Description

These quad gates are monolithic complementary MOS (CMOS) integrated circuits constructed with N- and P-channel enhancement mode transistors. They have equal source and sink current capabilities and conform to standard B series output drive. The devices also have buffered outputs which improve transfer characteristics by providing very high gain.

All inputs are protected against static discharge with diodes to V_{DD} and V_{SS}.

Features

- Low power TTL compatibility — Fan out of 2 driving 74L or 1 driving 74LS
- 5V–10V–15V parametric ratings
- Symmetrical output characteristics
- Maximum input leakage 1 μA at 15V over full temperature range

Schematic Diagrams

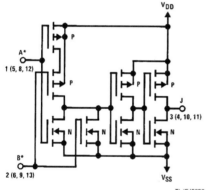

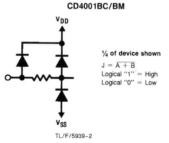

CD4001BC/BM

¼ of device shown
$J = \overline{A + B}$
Logical "1" = High
Logical "0" = Low

*All inputs protected by standard CMOS protection circuit.

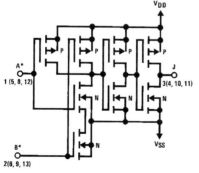

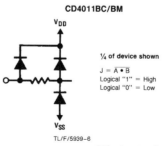

CD4011BC/BM

¼ of device shown
$J = \overline{A \cdot B}$
Logical "1" = High
Logical "0" = Low

*All inputs protected by standard CMOS protection circuit.

Absolute Maximum Ratings (Notes 1 and 2)

If Military/Aerospace specified devices are required, please contact the National Semiconductor Sales Office/Distributors for availability and specifications.

Voltage at any Pin	-0.5V to $V_{DD} + 0.5$V
Power Dissipation (P_D)	
Dual-In-Line	700 mW
Small Outline	500 mW
V_{DD} Range	$-0.5\ V_{DC}$ to $+18\ V_{DC}$
Storage Temperature (T_S)	$-65°C$ to $+150°C$
Lead Temperature (T_L)	
(Soldering, 10 seconds)	$260°C$

Operating Conditions

Operating Range (V_{DD})	3 V_{DC} to 15 V_{DC}
Operating Temperature Range	
CD4001BM, CD4011BM	$-55°C$ to $+125°C$
CD4001BC, CD4011BC	$-40°C$ to $+85°C$

DC Electrical Characteristics CD4001BM, CD4011BM (Note 2)

Symbol	Parameter	Conditions	$-55°C$ Min	$-55°C$ Max	$+25°C$ Min	$+25°C$ Typ	$+25°C$ Max	$+125°C$ Min	$+125°C$ Max	Units
I_{DD}	Quiescent Device Current	$V_{DD} = 5V, V_{IN} = V_{DD}$ or V_{SS}		0.25		0.004	0.25		7.5	μA
		$V_{DD} = 10V, V_{IN} = V_{DD}$ or V_{SS}		0.50		0.005	0.50		15	μA
		$V_{DD} = 15V, V_{IN} = V_{DD}$ or V_{SS}		1.0		0.006	1.0		30	μA
V_{OL}	Low Level Output Voltage	$V_{DD} = 5V$, $\|I_O\| < 1\ \mu A$		0.05		0	0.05		0.05	V
		$V_{DD} = 10V$		0.05		0	0.05		0.05	V
		$V_{DD} = 15V$		0.05		0	0.05		0.05	V
V_{OH}	High Level Output Voltage	$V_{DD} = 5V$, $\|I_O\| < 1\ \mu A$	4.95		4.95	5		4.95		V
		$V_{DD} = 10V$	9.95		9.95	10		9.95		V
		$V_{DD} = 15V$	14.95		14.95	15		14.95		V
V_{IL}	Low Level Input Voltage	$V_{DD} = 5V, V_O = 4.5V$		1.5		2	1.5		1.5	V
		$V_{DD} = 10V, V_O = 9.0V$		3.0		4	3.0		3.0	V
		$V_{DD} = 15V, V_O = 13.5V$		4.0		6	4.0		4.0	V
V_{IH}	High Level Input Voltage	$V_{DD} = 5V, V_O = 0.5V$	3.5		3.5	3		3.5		V
		$V_{DD} = 10V, V_O = 1.0V$	7.0		7.0	6		7.0		V
		$V_{DD} = 15V, V_O = 1.5V$	11.0		11.0	9		11.0		V
I_{OL}	Low Level Output Current (Note 3)	$V_{DD} = 5V, V_O = 0.4V$	0.64		0.51	0.88		0.36		mA
		$V_{DD} = 10V, V_O = 0.5V$	1.6		1.3	2.25		0.9		mA
		$V_{DD} = 15V, V_O = 1.5V$	4.2		3.4	8.8		2.4		mA
I_{OH}	High Level Output Current (Note 3)	$V_{DD} = 5V, V_O = 4.6V$	-0.64		-0.51	-0.88		-0.36		mA
		$V_{DD} = 10V, V_O = 9.5V$	-1.6		-1.3	-2.25		-0.9		mA
		$V_{DD} = 15V, V_O = 13.5V$	-4.2		-3.4	-8.8		-2.4		mA
I_{IN}	Input Current	$V_{DD} = 15V, V_{IN} = 0V$		-0.10		-10^{-5}	-0.10		-1.0	μA
		$V_{DD} = 15V, V_{IN} = 15V$		0.10		10^{-5}	0.10		1.0	μA

Connection Diagrams

CD4001BC/CD4001BM Dual-In-Line Package

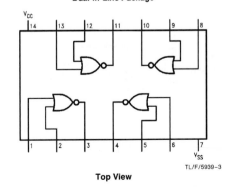

Top View

CD4011BC/CD4011BM Dual-In-Line Package

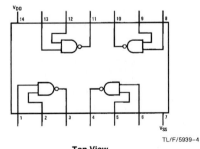

Top View

Order Number CD4001B or CD4011B

DC Electrical Characteristics CD4001BC, CD4011BC (Note 2)

Symbol	Parameter	Conditions	−40°C Min	−40°C Max	+25°C Min	+25°C Typ	+25°C Max	+85°C Min	+85°C Max	Units
I_{DD}	Quiescent Device Current	$V_{DD} = 5V$, $V_{IN} = V_{DD}$ or V_{SS}		1		0.004	1		7.5	μA
		$V_{DD} = 10V$, $V_{IN} = V_{DD}$ or V_{SS}		2		0.005	2		15	μA
		$V_{DD} = 15V$, $V_{IN} = V_{DD}$ or V_{SS}		4		0.006	4		30	μA
V_{OL}	Low Level Output Voltage	$V_{DD} = 5V$, $\|I_O\| < 1$ μA		0.05		0	0.05		0.05	V
		$V_{DD} = 10V$		0.05		0	0.05		0.05	V
		$V_{DD} = 15V$		0.05		0	0.05		0.05	V
V_{OH}	High Level Output Voltage	$V_{DD} = 5V$, $\|I_O\| < 1$ μA	4.95		4.95	5		4.95		V
		$V_{DD} = 10V$	9.95		9.95	10		9.95		V
		$V_{DD} = 15V$	14.95		14.95	15		14.95		V
V_{IL}	Low Level Input Voltage	$V_{DD} = 5V$, $V_O = 4.5V$		1.5		2	1.5		1.5	V
		$V_{DD} = 10V$, $V_O = 9.0V$		3.0		4	3.0		3.0	V
		$V_{DD} = 15V$, $V_O = 13.5V$		4.0		6	4.0		4.0	V
V_{IH}	High Level Input Voltage	$V_{DD} = 5V$, $V_O = 0.5V$	3.5		3.5	3		3.5		V
		$V_{DD} = 10V$, $V_O = 1.0V$	7.0		7.0	6		7.0		V
		$V_{DD} = 15V$, $V_O = 1.5V$	11.0		11.0	9		11.0		V
I_{OL}	Low Level Output Current (Note 3)	$V_{DD} = 5V$, $V_O = 0.4V$	0.52		0.44	0.88		0.36		mA
		$V_{DD} = 10V$, $V_O = 0.5V$	1.3		1.1	2.25		0.9		mA
		$V_{DD} = 15V$, $V_O = 1.5V$	3.6		3.0	8.8		2.4		mA
I_{OH}	High Level Output Current (Note 3)	$V_{DD} = 5V$, $V_O = 4.6V$	−0.52		−0.44	−0.88		−0.36		mA
		$V_{DD} = 10V$, $V_O = 9.5V$	−1.3		−1.1	−2.25		−0.9		mA
		$V_{DD} = 15V$, $V_O = 13.5V$	−3.6		−3.0	−8.8		−2.4		mA
I_{IN}	Input Current	$V_{DD} = 15V$, $V_{IN} = 0V$		−0.30		-10^{-5}	−0.30		−1.0	μA
		$V_{DD} = 15V$, $V_{IN} = 15V$		0.30		10^{-5}	0.30		1.0	μA

AC Electrical Characteristics* CD4001BC, CD4001BM

$T_A = 25°C$, Input t_r; $t_f = 20$ ns. $C_L = 50$ pF, $R_L = 200$k. Typical temperature coefficient is 0.3%/°C.

Symbol	Parameter	Conditions	Typ	Max	Units
t_{PHL}	Propagation Delay Time, High-to-Low Level	$V_{DD} = 5V$	120	250	ns
		$V_{DD} = 10V$	50	100	ns
		$V_{DD} = 15V$	35	70	ns
t_{PLH}	Propagation Delay Time, Low-to-High Level	$V_{DD} = 5V$	110	250	ns
		$V_{DD} = 10V$	50	100	ns
		$V_{DD} = 15V$	35	70	ns
t_{THL}, t_{TLH}	Transition Time	$V_{DD} = 5V$	90	200	ns
		$V_{DD} = 10V$	50	100	ns
		$V_{DD} = 15V$	40	80	ns
C_{IN}	Average Input Capacitance	Any Input	5	7.5	pF
C_{PD}	Power Dissipation Capacity	Any Gate	14		pF

*AC Parameters are guaranteed by DC correlated testing.

Note 1: "Absolute Maximum Ratings" are those values beyond which the safety of the device cannot be guaranteed. Except for "Operating Temperature Range" they are not meant to imply that the devices should be operated at these limits. The table of "Electrical Characteristics" provides conditions for actual device operation.

Note 2: All voltages measured with respect to V_{SS} unless otherwise specified.

Note 3: I_{OL} and I_{OH} are tested one output at a time.

National Semiconductor

November 1994

LM741 Operational Amplifier

General Description

The LM741 series are general purpose operational amplifiers which feature improved performance over industry standards like the LM709. They are direct, plug-in replacements for the 709C, LM201, MC1439 and 748 in most applications.

The amplifiers offer many features which make their application nearly foolproof: overload protection on the input and output, no latch-up when the common mode range is exceeded, as well as freedom from oscillations.

The LM741C/LM741E are identical to the LM741/LM741A except that the LM741C/LM741E have their performance guaranteed over a 0°C to +70°C temperature range, instead of −55°C to +125°C.

Schematic Diagram

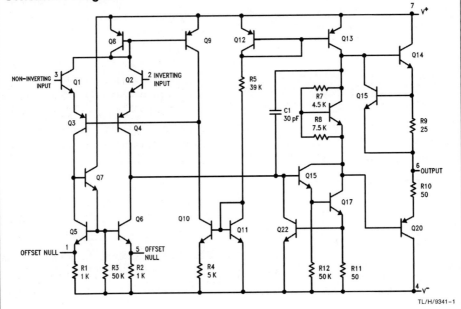

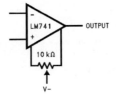

Offset Nulling Circuit

Absolute Maximum Ratings

If Military/Aerospace specified devices are required, please contact the National Semiconductor Sales Office/Distributors for availability and specifications.
(Note 5)

	LM741A	LM741E	LM741	LM741C
Supply Voltage	±22V	±22V	±22V	±18V
Power Dissipation (Note 1)	500 mW	500 mW	500 mW	500 mW
Differential Input Voltage	±30V	±30V	±30V	±30V
Input Voltage (Note 2)	±15V	±15V	±15V	±15V
Output Short Circuit Duration	Continuous	Continuous	Continuous	Continuous
Operating Temperature Range	−55°C to +125°C	0°C to +70°C	−55°C to +125°C	0°C to +70°C
Storage Temperature Range	−65°C to +150°C	−65°C to +150°C	−65°C to +150°C	−65°C to +150°C
Junction Temperature	150°C	100°C	150°C	100°C
Soldering Information				
N-Package (10 seconds)	260°C	260°C	260°C	260°C
J- or H-Package (10 seconds)	300°C	300°C	300°C	300°C
M-Package				
Vapor Phase (60 seconds)	215°C	215°C	215°C	215°C
Infrared (15 seconds)	215°C	215°C	215°C	215°C

See AN-450 "Surface Mounting Methods and Their Effect on Product Reliability" for other methods of soldering surface mount devices.

| ESD Tolerance (Note 6) | 400V | 400V | 400V | 400V |

Electrical Characteristics (Note 3)

Parameter	Conditions	LM741A/LM741E			LM741			LM741C			Units
		Min	Typ	Max	Min	Typ	Max	Min	Typ	Max	
Input Offset Voltage	$T_A = 25°C$										
	$R_S \leq 10\ k\Omega$					1.0	5.0		2.0	6.0	mV
	$R_S \leq 50\ \Omega$		0.8	3.0							mV
	$T_{AMIN} \leq T_A \leq T_{AMAX}$										
	$R_S \leq 50\ \Omega$			4.0							mV
	$R_S \leq 10\ k\Omega$						6.0			7.5	mV
Average Input Offset Voltage Drift				15							μV/°C
Input Offset Voltage Adjustment Range	$T_A = 25°C, V_S = ±20V$	±10				±15			±15		mV
Input Offset Current	$T_A = 25°C$		3.0	30		20	200		20	200	nA
	$T_{AMIN} \leq T_A \leq T_{AMAX}$			70		85	500			300	nA
Average Input Offset Current Drift				0.5							nA/°C
Input Bias Current	$T_A = 25°C$		30	80		80	500		80	500	nA
	$T_{AMIN} \leq T_A \leq T_{AMAX}$			0.210			1.5			0.8	μA
Input Resistance	$T_A = 25°C, V_S = ±20V$	1.0	6.0		0.3	2.0		0.3	2.0		MΩ
	$T_{AMIN} \leq T_A \leq T_{AMAX}$, $V_S = ±20V$	0.5									MΩ
Input Voltage Range	$T_A = 25°C$							±12	±13		V
	$T_{AMIN} \leq T_A \leq T_{AMAX}$				±12	±13					V
Large Signal Voltage Gain	$T_A = 25°C, R_L \geq 2\ k\Omega$										
	$V_S = ±20V, V_O = ±15V$	50									V/mV
	$V_S = ±15V, V_O = ±10V$				50	200		20	200		V/mV
	$T_{AMIN} \leq T_A \leq T_{AMAX}$, $R_L \geq 2\ k\Omega$,										
	$V_S = ±20V, V_O = ±15V$	32									V/mV
	$V_S = ±15V, V_O = ±10V$					25			15		V/mV
	$V_S = ±5V, V_O = ±2V$	10									V/mV

Electrical Characteristics (Note 3) (Continued)

Parameter	Conditions	LM741A/LM741E Min	LM741A/LM741E Typ	LM741A/LM741E Max	LM741 Min	LM741 Typ	LM741 Max	LM741C Min	LM741C Typ	LM741C Max	Units
Output Voltage Swing	$V_S = \pm 20V$ $R_L \geq 10 k\Omega$ $R_L \geq 2 k\Omega$	± 16 ± 15									V V
	$V_S = \pm 15V$ $R_L \geq 10 k\Omega$ $R_L \geq 2 k\Omega$				± 12 ± 10	± 14 ± 13		± 12 ± 10	± 14 ± 13		V V
Output Short Circuit Current	$T_A = 25°C$ $T_{AMIN} \leq T_A \leq T_{AMAX}$	10 10	25	35 40		25			25		mA mA
Common-Mode Rejection Ratio	$T_{AMIN} \leq T_A \leq T_{AMAX}$ $R_S \leq 10 k\Omega, V_{CM} = \pm 12V$ $R_S \leq 50\Omega, V_{CM} = \pm 12V$	80	95		70	90		70	90		dB dB
Supply Voltage Rejection Ratio	$T_{AMIN} \leq T_A \leq T_{AMAX}$, $V_S = \pm 20V$ to $V_S = \pm 5V$ $R_S \leq 50\Omega$ $R_S \leq 10 k\Omega$	86	96		77	96		77	96		dB dB
Transient Response Rise Time Overshoot	$T_A = 25°C$, Unity Gain		0.25 6.0	0.8 20		0.3 5			0.3 5		μs %
Bandwidth (Note 4)	$T_A = 25°C$	0.437	1.5								MHz
Slew Rate	$T_A = 25°C$, Unity Gain	0.3	0.7			0.5			0.5		$V/\mu s$
Supply Current	$T_A = 25°C$					1.7	2.8		1.7	2.8	mA
Power Consumption	$T_A = 25°C$ $V_S = \pm 20V$ $V_S = \pm 15V$		80	150		50	85		50	85	mW mW
LM741A	$V_S = \pm 20V$ $T_A = T_{AMIN}$ $T_A = T_{AMAX}$			165 135							mW mW
LM741E	$V_S = \pm 20V$ $T_A = T_{AMIN}$ $T_A = T_{AMAX}$			150 150							mW mW
LM741	$V_S = \pm 15V$ $T_A = T_{AMIN}$ $T_A = T_{AMAX}$					60 45	100 75				mW mW

Note 1: For operation at elevated temperatures, these devices must be derated based on thermal resistance, and T_j max. (listed under "Absolute Maximum Ratings"). $T_j = T_A + (\theta_{jA} P_D)$.

Thermal Resistance	Cerdip (J)	DIP (N)	HO8 (H)	SO-8 (M)
θ_{jA} (Junction to Ambient)	100°C/W	100°C/W	170°C/W	195°C/W
θ_{jC} (Junction to Case)	N/A	N/A	25°C/W	N/A

Note 2: For supply voltages less than $\pm 15V$, the absolute maximum input voltage is equal to the supply voltage.

Note 3: Unless otherwise specified, these specifications apply for $V_S = \pm 15V$, $-55°C \leq T_A \leq +125°C$ (LM741/LM741A). For the LM741C/LM741E, these specifications are limited to $0°C \leq T_A \leq +70°C$.

Note 4: Calculated value from: BW (MHz) = 0.35/Rise Time(μs).

Note 5: For military specifications see RETS741X for LM741 and RETS741AX for LM741A.

Note 6: Human body model, 1.5 kΩ in series with 100 pF.

Appendix E Design Project Case Study

This appendix contains a highly condensed case study of a capstone design project that was developed by a team of students utilizing the principles of this textbook. The complete project report can be downloaded from *www.mhhe.com/fordcoulston* along with a number of other case studies.

The Visual Aid
Ryan Andrus, Luis Catoni, Carl Schnur, and Freddy Chiu
A Senior Project Report Submitted to the Faculty of
Electrical, Computer, and Software Engineering
Penn State Erie, The Behrend College
April 2006

1 Problem Statement

1.1 Need
Visually impaired people often have mobility difficulties due to limited spatial sensing—determining where objects are. Although many receive education in mobility techniques and enhance other senses to create better awareness of surroundings, there is a need for a more accurate spatial description. For example, according to the Vision and Blindness Resources Center in Erie, visually impaired people are able to detect walls because of different sounds in the environment. However, a tree branch, stairs, or a street sign may be undetected. As a result of objects that cannot be perceived by sensorial means, many devices have been designed to detect objects in the path of a visually impaired individual. Common mobility resources are guide dogs, canes, and electronic travel aid (ETA) devices. Unfortunately, these resources are either limited or too expensive. There is a need for a device to provide an effective way to detect objects and be cost-efficient.

1.2 Objective
The goal is to design and implement a digital system that gives visually impaired an enhanced awareness of their surroundings. The system will detect objects and provide real-time feedback to the user according to the size, position, and distance of the object.

1.3 Research Survey
For over 30 years, people have been attempting to invent an electronic device to help the blind navigate. According to the American Foundation for the Blind [1], there are about 10 million blind and visually impaired individuals in the United States.

According to The Seeing Eye [2], an organization committed to train dogs for the blind, around 1% of the visually impaired use guide dogs. Many do not choose this option because of allergies, facilities needed for dog care, training for both the dog and the individual, and, finally, personal preference. The rest of the blind community relies on canes, electronic travel aids (ETAs), and their perceptual senses.

The cane is the most popular tool for object detection for the blind. Nevertheless, this tool is limited, as it is difficult to locate objects that are not at the ground level. For example, a tree branch will go undetected by an individual using a cane. However, there are ETAs designed to overcome this problem. The most popular ETA is a cane that contains built-in laser sensors for object detection called the LaserCane [3]. This system integrates laser sensors near the handle of the cane and detects objects in three different angles. This system provides feedback to the user by sound or by vibrations produced on the side of the cane sensed by touch. Although this design offers the blind more resources to identify possible objects in front, it has limited feedback. For instance, the laser sensors fire a beam so narrow that users must master the movements of the cane in order to detect objects accurately. Moreover, this device costs around $2500, which places an attainability issue for many visually impaired.

According to the director of the Vision and Blindness Resources Center in Erie—who was interviewed as part of the project's research—the visually impaired are trained to orient themselves and navigate by using their senses. However, major difficulties arise from objects that cannot be sensed. Therefore, basic mobility needs of visually impaired consist of object detection with a range of approximately 1 meter from the body, as a minimum, and 2 feet wide—the width of the body. Such objects may include desks, chairs, steps, and tree branches. In an interview, a visually impaired student on campus who uses a guide dog said that surface differences of approximately 2 inches should be considered dangerous, and must be detected by the ETA system.

To meet the needs and common hazards that the visually impaired encounter, key characteristics must be carefully addressed when designing an ETA device. First, the sensing system must be able to detect objects of varied sizes and cover a wide range for ease of detection. Second, GPS and compasses may be used for orientation. Third, an intuitive feedback system must be used. Such a system cannot employ sound, as this interferes with the hearing of the visually impaired. Finally, the system must be affordable. Research shows that no single device meets all these needs, and none has been widely accepted by the blind community.

1.4 User Needs and Objective Tree

The team interviewed a number of sources to identify the objectives in the tree in Figure E.1.

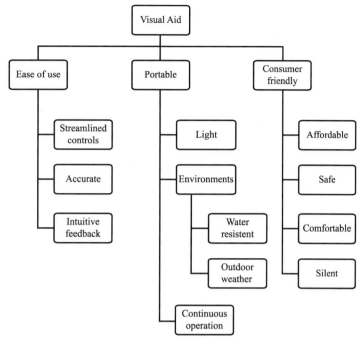

Figure E.1 The needs and objective tree.

Weightings of the objectives were determined by using the pairwise comparison method for all levels of the objective tree. The results for the highest level of the tree are portability (0.20), consumer friendly (0.30), and ease of use (0.50). This result arises from the need that the ETA must be easy to use, because an overly complex interface would create a high barrier to using and then accepting the device. Along the same lines, the device should create a minimum amount of disturbance to the user's environment. Finally, while important, portability is the least of the three concerns.

2 The Requirements Specification

Table E.1 Requirements Specification for the Visual Aid.

Marketing Requirements	Engineering Requirements	Justification
1, 2, 10, 13	1. The system's total weight will not exceed 5 lb.	Based on the weight of other portable devices, such as laptops, and the weight of book bags.
7, 8, 13	2. The system will have a single control to turn it on/off.	Based on the user's need for the system to be as simple and intuitive as possible and the average person's familiarity with technology.

5, 13	3.	The system will operate on full charge for at least 3 hours.	Based on the expected daily use by considering average daily walking time (2.1 miles at 3.3 MPH ≈ 38 minutes/day) by a factor of approximately 5.
6	4.	The system should not exceed $600.	Based on the cost of competing products [6], such as the LaserCane.
11	5.	Uneven surfaces (steps, rocks) of at least ±1.25 in high will be detected at a distance of at least 3 ft away from the user.	Based on the size of possible hazards, such as rocks, bottles, sign posts, and stairs.
11, 14	6.	The system will detect objects that are at least 1 in wide and 2 in high at a distance of at least 3 ft from the user and as far as 7 ft in an area 2 ft wide, and provide sensory feedback. Also, the system should detect street signs and posts.	Based on the user's need to be notified of hazards in a timely and intuitive way.
12	7.	The system should not produce noise exceeding 40 dB.	Based on comparisons of sound levels in office, bedroom, and living room environments [7].
3	8.	The system will be built with components that can operate in temperatures ranging from 0°F to 120°F.	Based on the expected range of temperatures the user might operate the device in.
3, 4, 9	9.	The system components will be enclosed to be water-resistant.	Based on outdoor environments and conditions (snow, rain, humidity) the user might face and the need for the user to be protected from electrical shock.
14	10.	The system will refresh its output according to sensor readings 5 times per second.	Based on the average reaction time of humans (0.33 seconds) [8].

Marketing Requirements
(1) Portability, (2) weight, (3) operable environments, (4) water resistant, (5) operation time, (6) cost, (7) ease of use, (8) streamlined controls, (9) safety, (10) comfortable, (11) accurate, (12) noise level, (13) consumer-friendly, (14) intuitive feedback

3 The Design

The Visual Aid device has three major physical components: (1) a set of sensors, (2) a matrix of vibrating motors, and (3) a processing unit. These physical components are interconnected through the processing unit and are attached to the user as in Figure E.2 below.

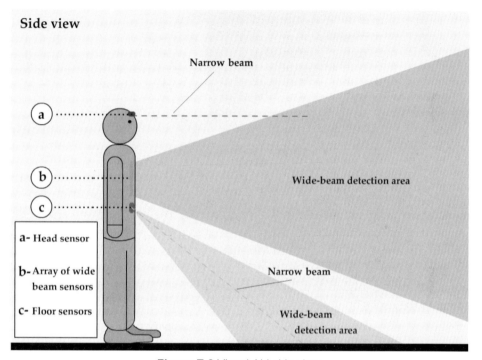

Figure E.2 Visual Aid side view.

The set of sensors is distributed in three modules. The head module consists of a narrow-beam sensor (represented as a dashed line) mounted in the head of the user. The purpose of this module is to provide the user with accurate information regarding an objects position, size, and distance.

The second module is the array of six wide-beam sensors (Figure E.3). The use of wide-beam sensors enables the system to detect objects in a larger area. In order to give more detailed information on an object's spatial location the sensors are arranged such that portions of their beams overlap with each other. The sensor matrix is aligned to give a wider range in the vertical orientation than the horizontal.

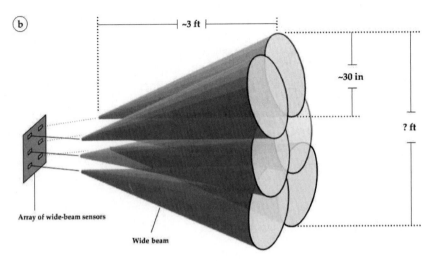

Figure E.3 Wide-beam sensor array.

The third module detects objects and steps up or down at the ground level. This module consists of a wide-beam sensor and a narrow-beam sensor. The wide-beam sensor detects objects and obstructions on the ground. The narrow-beam sensor detects uneven surfaces such as steps and stairs. This sensor will be positioned as shown in Figure E.4 with an angle θ. This angle is adjustable to fit user needs. On flat surfaces, the sensor will detect a distance $x + z$. When a step is encountered, the distance read by the sensor will be diminished to x. In order to avoid "false" detection of steps due to the movement of the user's body while walking, a digital filter is used.

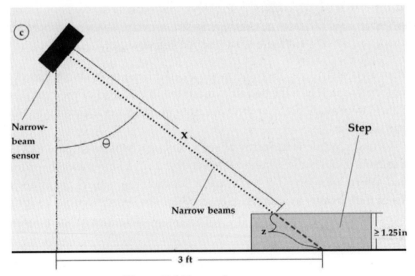

Figure E.4 Narrow-beam sensor.

The second major component of the system consists of two motors located in the front area of the user (see Figure E.5), and 12 vibrating motors configured in a 3 × 3 matrix on the back of the user (see Figure E.6). The head unit motor allows the user to scan for objects by synchronizing movement of these sensors to those of the head; the motor will generate higher intensity for closer objects and lower intensity for farther objects. The narrow-beam motor is used to provide feedback about an uneven surface detected by the narrow beam sensor shown in Figure E.4. The 3 × 3 matrix shown in Figure E.6 is mounted on the user's back and provides feedback according to objects detected in front of the users. For example, if an object is detected in the lower right corner, then the lower right motor will vibrate. The matrix motors also vary their intensity according to the distance of objects detected.

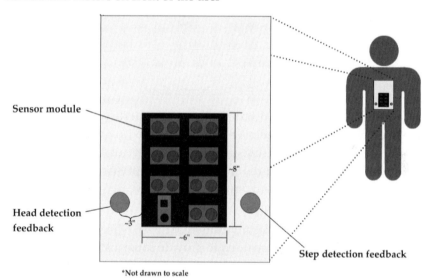

Figure E.5 Sensors and front motors.

Motors on the back of the user

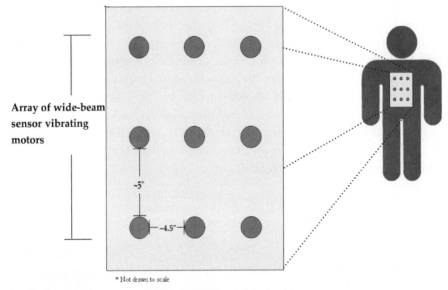

Figure E.6 The motors matrix.

The array of motors is mounted on the user with a vest. The array of sensors is located on the user's chest and the motor matrix is located on the user's back. The last physical component is the processing unit. This consists of the microcontroller unit, which will read the inputs from all the sensors and will provide corresponding outputs to the user by controlling the vibrating motors.

3.1 The Functional Decomposition

The following sections outline the decomposition of Level 0 diagram into a Level 1 diagram with the design rationale included to provide some justification for the decisions made.

Level 0

The Level 0 description of the Visual Aid covers the overall input, output, and behavior expected of the device (Figure E.7).

Level 0

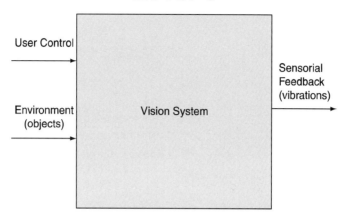

Figure E.7 Level 0.

The rationale for our Level 0 design was to keep the system easy to control by a nontechnical user. Too many buttons would confuse the user. The system will take environmental data on the surroundings (objects) as input and provide vibration feedback as output to the user. The only control is an on/off switch, with everything else automated. The object is to provide the user with carefree operation.

Module	Vision System
Inputs	- User control: on/off - Environment (objects): objects at least 1 in wide and 2 in high found in a 2-ft-wide area in front of the user, including steps.
Outputs	- Sensorial feedback (vibrations): a 3 × 3 matrix of vibrating motors, plus three additional vibrating motors.
Functionality	Alerts the user intuitively by means of vibrating motors as to what area and distance in front of the user objects are and/or if there is a step. Detected objects are at least 1 in wide and 2 in high at a distance of at least 3 ft from the user and as far as 7 ft in an area 2 ft wide. Variations of height (steps, rocks) of at least ±1.25 in are detected at a distance of at least 3 ft away from the user.

Level 1

The Level 1 diagram (Figure E.8) reveals what is inside the Level 0 box shown in Figure E.7.

Level 1

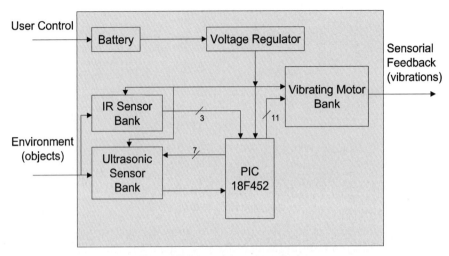

Figure E.8 Level 1 system design.

There are four main components in this part of our system design: the IR sensor bank, ultrasonic sensor bank, PIC18F452, and vibrating motor bank. Also, there are modules for supplying power to the components such as a voltage regulator.

The reason for breaking the design into these four main components is because each one provides a separate and distinct functionality toward the overall operation of the system. The IR sensor bank is responsible for detecting changes in elevation of the surface (steps, curbs, ledges) and also for identifying objects targeted by the user's head. The ultrasonic sensor bank is responsible for detecting objects directly in front of the user and also indicating the object's general location. The PIC18F452 is the control unit of the entire system. It will read data from the sensor banks, process the data, and provide feedback control via the vibrating motor bank. Finally, the vibrating motor bank, which is mounted on the back of the user, provides sensorial feedback to the user in the form of vibrations of varying intensity based on the proximity of the object and relative to the location of the object.

Module	Battery
Inputs	- User control: on/off
Outputs	- 9.6 V DC
Functionality	Provide power to all electronics in the system.

Module	Voltage Regulator
Inputs	- 9.6 V DC
Outputs	- 5 V DC with up to 1 A of current.
Functionality	Convert the battery's 9.6 V DC to 5 V DC.

Module	IR Sensor Bank
Inputs	- 5 V DC for power. - Environment (objects): objects at head level and steps.
Outputs	- V_0–V_2: voltages ranging from 0.55 V DC to 2.8 V DC
Functionality	One sensor to detect objects at head level and another sensor angled to detect uneven surfaces.

Module	Ultrasonic Sensor Bank
Inputs	- 5 V DC for power. - Environment (objects): objects at least 1 in wide and 2 in high found in a 2-ft-wide area in front of the user. - D_0–D_7: forwarded 5-V DC 10-μs control signal from PIC to selected sensor.
Outputs	- U_0–U_7: 0–5 V DC pulse signal from 100 μs to 18 ms corresponding to selected sensor.
Functionality	Detects objects on the floor and in front of the user (under head level), and gives feedback regarding location and distance to object.

Module	Vibrating Motor Bank
Inputs	- 5 V DC for power. - M_0–M_{11}: PWM signals from 0–5 V DC
Outputs	- Sensorial feedback (vibrations): sustained or pulsed vibrations with varied intensities.
Functionality	Converts the processed sensor information into feedback for the user.

Module	PIC 18F452
Inputs	- 5 V DC for power. - R_0–R_2: analog voltage feedback from 0.55 V to 2.8 V - U_X: 0–5 V DC pulse signal from 100 μs to 18 ms.
Outputs	- S_0–S_2: digital select lines - U_C: 5 V DC 10 μs pulse - M_0–M_{11}: PWM signals with ? period and ?% duty cycle
Functionality	Gather data from ultrasonic and IR sensors, then convert it to PWM signals that drive the vibrating motors.

The first decision for the design was to determine which method was best to detect objects. Table E.2 outlines some of the advantages and disadvantages of a few methods.

Table E.2 Sensor selection decision.

Method	Advantages	Disadvantages
Ultrasonic sensor	Wide range Inexpensive	Less accurate
IR sensor	Accurate Inexpensive	Limited scope and range
Lasers	Accurate Long-distance detection	Expensive Limited scope
Radar	Infinite range Near perfect accuracy	Very expensive Large and heavy Expert knowledge required

Ultrasonic sensors were capable of detecting the area that we required (range: 1 in to 10 ft, with area of detection growing wider as distance increases). Infrared sensors were chosen because they have a relatively narrow beam, which can be used to easily pinpoint the location of an object and also to conduct precise measurements to changes in elevation.

The main disadvantages to laser and radar technology were high costs and bulkiness. Also, no group member is familiar with the technology and large amounts of time to learn would be required. Laser technology is useful in detecting small objects at large distances, but we found that the large range of the lasers was unnecessary and the price of this technology is too high.

The next decision falls in the category of feedback to the user. Table E.3 outlines the advantages and disadvantages of two methods.

Table E.3 User feedback selection.

Method	Advantages	Disadvantages
Vibration	Quiet operation Inexpensive	May take time for user to get used to this type of sensorial feedback
Audible	Unlimited capability	Interferes with a blind user's vital hearing ability

The two alternatives to choose from involving feedback to the user were either through vibrations or audible sounds. After some interviews and extensive research of the visually impaired, we learned that audible feedback would interfere with their vital sense of hearing. Most blind people are trained to enhance their hearing capability to navigate naturally as they walk so they typically avoid using devices that produce audible feedback. We chose to use the vibration method because of its quiet operation, although this method will likely require some getting used to as people are not used to having things producing this type of sensorial feedback on their body.

Finally, we needed to choose a method to control the other three components of the design. We chose the PIC18F452 because it provides subsystems that are very useful for our design such as multiple timer modules and an ADC subsystem. Also, our familiarity with this device encouraged us to use it for our main control. The PIC also provides many I/O ports and plenty of internal flash memory and is an inexpensive device.

3.2 High-Level Software Description

The development of the software system exhibits the same modularity as the hardware design shown in Figure E.8. Figure E.9 shows the overall behavior of the system as a state diagram, listing the sequence of operations that are performed by the MCU.

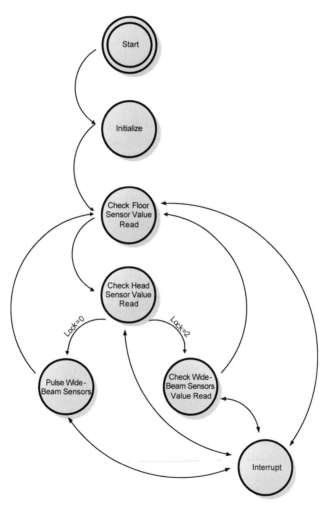

Figure E.9 Main state diagram for software.

In the list that follows, each of the states in Figure E.9 is described in more detail regarding its functionality.

- **Start**—This is a dummy state to represent the system when it is turned on. It will transition to Initialize Hardware state.
- **Initialize**—This is a function that runs at the beginning of the program (upon power-up of the system) to initialize the different hardware and modules used.
- **Check Floor Sensor Value Read**—The values read from the floor unit, specifically the narrow-beam sensor, are used to determine whether an uneven surface has been detected or not. If the surface is uneven, activate the corresponding motor.
- **Check Head Sensor Value Read**—The values read from the head unit are used to determine and set intensity at which the motor corresponding to this unit should operate.
- **Pulse Wide-Beam Sensors**—This calls a function to pulse the ultrasonic sensors.
- **Check Wide-Beam Sensors Value Read**—The values from the ultrasonic sensors are used to determine where objects are and to activate the corresponding motors.
- **Interrupt**—This is a state to represent the various interrupts that can occur during the program. The only state where the interrupts are deactivated is in the Check Wide-Beam Sensors Value Read state. They have been deactivated because of sensitivity in the program to determine the pattern in which motors should be vibrating.

4 Design Verification and Testing

The design testing is presented in the order that the tests were performed, from unit to acceptance tests.

4.1 Unit Testing

The performance of the IR sensor was critical to the success of the system, as it would be observing the floor and informing the user of any upcoming steps. We had to test the sensor to see if it could distinguish the height bob of a walker from the drop-off of a set of steps (Table E.4).

Table E.4 IR sensor test.

Test Name: IR Sensor Test			Test Number: 101		
Test Description: Verify that the IR sensor is correctly reading distance without unnecessary noise.					
Test Information					
Name of Tester: Luis		Date: 3/18/06		Time: 2:00 PM	
#	Procedure	Pass	Fail	N/A	Comments
1	Mount the IR sensor and hold it vertically at waist level.	x			
2	Start recording data while walking along a flat, obstacle-free path.	x			

3	Analyze that data to ensure that minimal noise is created from the up and down movement of walking.	x			Used Luis, Carl, and a rolling cart.
4	Repeat test with stairs along the walking path.	x			
5	Analyze the values to ensure that they differ enough from data along the flat surface.	x			See data in Table E.9 for plots with different filters used.

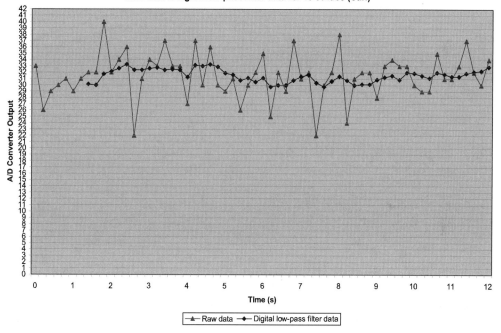

Figure E.10 ADC samples from the narrow-beam sensor, raw and filtered.

The graph in Figure E.10 shows the effects of using a digital low-pass filter on the readings from an IR sensor mounted at stomach level on a walking person. The person walked on an even surface in a laboratory environment to gather the data samples. In comparison with the data obtained from the cart, the raw data has more noise, yet the low-pass filter manages to stabilize the value considerably.

4.1.1 Integration Testing

The integration test (Table E.5) checks that individual ultrasonic sensors could trigger individual motors. This is critical to ensure that the system gives proper feedback to the user about the environment.

Table E.5 Integration testing

Test Name: Ultrasonic Sensor/Motor Test				Test Number: 200	
Test Description:					
Test Information					
Name of Tester: Freddy and Ryan		Date: 3/25/06		Time: 1:00 PM	
#	Procedure	Pass	Fail	N/A	Comments
1	The tester should be wearing the visual aid system with the ultrasonic sensors and vibrating motors properly installed.	x			
2	Have someone place an object in the path of each individual sensor.	x			
3	Verify that the user is able to identify which position on the vibrating motor bank is being activated. Also, ensure the correct motor is being activated.	x			
4	Repeat step (3) for each ultrasonic sensor combination in order to activate each vibrating motor separately.	x			
5	Next, repeat the same procedure but activate multiple ultrasonic sensors and vibrating motors simultaneously and ensure that the unit is functioning properly.	x			See conclusions below.

This testing was performed many times and helped us to determine the proper adjustments/angles of the sensors. Performing these tests also gave us insight on what to set the sensor thresholds at in the software.

4.1.2 Acceptance Testing

One of the main engineering requirements is: *The system will detect objects that are at least 1 in wide and 2 in high at a distance of at least 3 ft from the user and as far as 7 ft in an area 2 ft wide, and provide sensory feedback. Also, the system should detect street signs and posts.* In order to ensure that this requirement was met, the acceptance test in Table E.6 was constructed.

Table E.6 Surface test.

Test Name: Object Detection Test				Test Number: 305
Test Description: Verify that the system can detect objects that are at least 1 in wide and 2 in high at a distance of at least 3 ft from the user and as far as 7 ft in an area 2 ft wide, and provide sensory feedback. Also, the system should detect street signs and posts				
Test Information				
Name of Tester: Freddy and Ryan		Date: 3/18/06		Time: 12:30 PM
#	Procedure	Pass	Fail	N/A	Comments
1	On a flat level surface, first mark a spot where the system will take readings from.	x			
2	Next, mark out a rectangular box on the floor, starting 3 ft from the testing position. The box should be centered with the testing spot and be 2 ft wide and 4 ft long. This box is used as reference for placing objects. (See comments)	x			2', 4', 3'
3	Place a small object which is approximately 1" wide and 2" high across the front line (left corner, right corner, and middle). Verify that the object is detected at each position and the appropriate motors are activated.	x			x x x
4	Now, repeat step (3) but place the object across the back line.	x			x x x
5	Again, repeat step (3) but place the object across the center of the box.	x			x x x
6	Record all results and repeat test with objects of different sizes (poles, chairs, etc.) to ensure proper detection.	x			To verify this test we set up an obstacle course with objects that the user had to navigate hrough. See video for results of this test.

4.2 Requirements Verification

Tests were constructed and run for each of the engineering requirements. These are enumerated in Table E.7.

Table E.7 Requirements verification reference.

Engineering Requirement	Test Verification
The system's total weight will not exceed 5 lb.	Showed that weight did not exceed 5 lb. Test: Weight Test Test #: 300
The system will have a single control to turn it on/off.	Showed that on/off switch works properly. (Sometimes needs to be switched twice) Test: On/Off Switch Test Test #: 301
The system will operate on full charge for at least 3 hours.	Showed that the system will operate for at least 3 hours under normal conditions. Test: Full Charge Operation Time Test #: 302
The system cost should not exceed $600.	Showed that the total expenses to construct the system did not exceed $600 budget. Test: Budget Test Test #: 303
Uneven surfaces (steps, rocks) of at least ±1.25 in high are detected at a distance of at least 3 ft away from the user.	Showed that the system can properly detect uneven surfaces. Test: Uneven Surface Test Test #: 304
The system will detect objects that are at least 1 in wide and 2 in high at a distance of at least 3 ft from the user and as far as 7 ft in an area 2 ft wide, and provide sensory feedback. Also, the system should detect street signs and posts.	Showed that the system can properly detect objects of a minimum size within the boundaries of the test area. Test: Object Detection Test Test #: 305
The system should not produce noise exceeding 40 dB.	Showed that system noise is < 40 dB and is not distracting during quiet conversation. Test: Sound Level Test Test #: 306
The system is built with components that can operate in temperatures ranging from 0°F to 120°F.	Showed that all components of system will operate properly with the temperature range. Test: Operating Environment Temperature Test #: 307
The system components are enclosed to be water-resistant.	Components are not sealed in our prototype. Could be made water-resistant and tested. Test: Water-Resistance Test Test #: 308
The system will refresh its output according to sensor readings 5 times per second.	Showed that the system will refresh its output at least 5 times per second. Test: Refreshing System Output Test Test #: 309

5 Summary and Conclusions

Overall, the project was completed successfully and all the critical requirements were met. We had the chance to meet with a visually impaired student and have him test the system. We were content with the satisfaction he demonstrated toward the system. Considering this system is at the prototype level, we are happy with the level of functionality the system was able to reach.

It was a challenge, but we were able to have the ultrasonic sensors and IR sensors work together without having to increase the complexity level of the system. This is important because the functions of both sensors are critical; the ultrasonic sensors perform the bulk of the detection work, and one of the IR sensors is used to detect steps.

While the system works well, there are some possible improvements to note:

- Place the motors on the front in a more sensitive area of the user
- Make the system adaptable (turn off motors after long periods of time)
- Design a vest better suited for everyday use
- Use a faster crystal or microcontroller to improve refresh rate
- Replace the IR sensors with more reliable and consistent sensors
- Put the circuitry in a PCB to reduce space and weight
- Fine-tune object detection with the ultrasonic sensors

The experience of working as a design team has taught us valuable lessons. We learned the importance of communicating as a team, and understanding the abilities and limitations of each member in order to maximize the team's performance. Also, we realized how much of a difference a good planning can make; having a good design from the previous semester made it a lot easier to jump into work at the beginning of the term.

On a more technical level, we learned how much the use of analog and digital filters can help in cleaning signal noise. This played a major role in having the system operate correctly.

Finally, meeting with a blind person made us realize how important and useful it is to obtain feedback from the expected users of the system.

6 References

[1] American Foundation for the Blind, "Blindness Statistics," February 2006, http://www.afb.org.
[2] The Seeing Eye Inc., "About Us," February 2006, http://www.seeingeye.org/AboutUs.asp.
[3] Nurion-Raycal, "LaserCane N-2000, Electronic Travel Aids," February 2006.
[4] GDP Research, "The Miniguide, an ultrasonic mobility aid, electronic travel aid (ETA)," February 2006, http://www.gdp-research.com.au/minig_1.htm.

[5] Robotron Group, "Robotron Sensory Tools—C2 Talking Compass," February 2006, http://www.sensorytools.com/c2.htm.
[6] D. His Yen (yen@noogenesis.com), "Currently Available Electronic Travel Aids," September 2005, http://www.noogenesis.com/eta/current.html.
[7] F. Miyara (fmiyara@fceia.unr.edu.ar), "Sound Levels," February 2006, http://www.eie.fceia.unr.edu.ar/~acustica/comite/soundlev.htm.
[8] Division of Educational Programs, Argonne National Laboratory, "Cell Phone and Reaction Time," September 2002, http://www.newton.dep.anl.gov/askasci/gen01/gen01264.htm.
[9] Hong Z. Tan. "Haptic Interfaces," February 8, 2006. http://dynamo.ecn.purdue.edu/~hongtan/.

References

ABC01	ABC News Productions Nightline, "The Deep Dive," Video N990713-51, dated July 13, 1999, copyright 2001.
ABE03	*Criteria for Accrediting Engineering Programs*, ABET Engineering Accreditation Commission, 2003.
Abl91	R. Abler, "Good Product Designs Start with Good Specifications," *Medical Device Diagnostic Industry*, pp. 156–157, June 1991.
Ada01	J. L. Adams, *Conceptual Blockbusting*, 4th ed., Perseus Publishing, 2001.
Alb07	L. Al-Busaidi, M. Bellavia, and E. Roseborough, "iPod^{TM} Hands Free Device," *Penn State Behrend Electrical and Computer Engineering Senior Design Report*, 2006.
Alt99	G. Altshuller, *The Innovation Algorithm: TRIZ, Systematic Innovation, and Technical Creativity*, Technical Innovation Center, 1999.
And07	N. Andre, J. Kolb, and J. Thaler, "Portable Aerial Surveillance System," *Penn State Behrend Electrical and Computer Engineering Senior Design Report*, 2006.
Ang00	R. B. Angus, N. R. Gunderson, and T. P. Cullinane, Planning, Performing and Controlling Projects: Principles and Applications, 2nd ed., Prentice Hall, 2000.
App01	D. Apple and K. Krumsieg, *Pacific Crest Teaching Institute Handbook*, 2001.
Bai81	J. E. Baird Jr., "How to Overcome Errors in Public Speaking," IEEE Transactions on Professional Communication, vol. PC-24, no. 2, pp. 94–98, 1981.
Bee03	D. F. Beer, editor, Writing and Speaking in the Technology Professions: A Practical Guide, Wiley-IEEE Press, 2003.
Bel94	L. Bellamy, D. Evans, D. Linder, B. McNeill, and G. Raupp, Team Training Workbook, Arizona State University, 1994. Available at ww.eas.asu.edu/~asufc/teaminginfo/teamwkbk.pdf.
Ben01	S. Bennett and J. Skelton, Schaum's Outline of UML, McGraw-Hill, 2001.
Bur03	E. M. Burke and B. M. Coyner, Java Extreme Programming Cookbook, O'Reilly, 2003.
Cag02	J. Cagan and C. M. Vogel, *Creating Breakthrough Products: Innovation from Product Planning to Program Approval*, Financial Times/Prentice Hall, 2002.
Car89	G. Carle, "Handling a Hostile Audience—With Your Eyes," IEEE Transactions on Professional Communication, vol. 32, no. 1, pp. 29–31, 1989.

Cha71   N. Chapin, Flowcharts, Auerbach, 1971.

Cha02   K. Chakraborty and P. Mazumder, *Fault-Tolerance and Reliability Techniques for High-Density Random-Access Memories*, Prentice Hall PTR, 2002.

Cha02b   T. Chambers, "A Piercian Approach to Professional Ethics Instruction," *IEEE Transactions on Professional Communication*, vol. 45, no. 1, pp. 45–49, March, 2002.

Cro00   N. Cross, Engineering Design Methods: Strategies for Product Design, 3rd ed., John Wiley & Sons, 2000.

Deb67   E. DeBono, *New Think: The Use of Lateral Thinking in the Generation of New Ideas*, New York, Basic Books, 1967.

Deb70   E. DeBono, *Lateral Thinking: Creativity Step by Step*, Harper and Row, 1970.

Dec84   B. Decker, "A Good Speech Is Worth a Thousand Words," IEEE Transactions on Professional Communication, vol. PC-27, no. 1, pp. 32–34, March, 1984.

Deg81   R. T. DeGeorge, "Ethical Responsibilities of Engineers in Large Organizations: The Pinto Case," *Business and Professional Ethics Journal*, vol. 1, no. 1, pp. 1–14, 1981.

Del71   A. L. Delbecq and A. H. VandeVen, "A Group Process Model for Problem Identification and Program Planning," *Journal of Applied Behavioral Science VII*, pp. 466–491, July/August, 1971.

Die00   G. E. Dieter, *Engineering Design*, 3rd ed., McGraw-Hill, 2000.

Dom01   P. G. Dominick et al., *Tools and Tactics of Design*, John Wiley & Sons, 2001.

Dym94   C. Dym, *Engineering Design: A Synthesis of Views*, Cambridge University Press, 1994.

Elm94   R. Elmasri and S. B. Navathe, Fundamentals of Database Systems, Cummings, 1994.

Ern97   R. Ernst, *Embedded System Architectures in Hardware/Software Co-Design: Principles and Practice*, Kluwer Academic Publishers, 1997.

Ese03   T. Esek, J. Hunt, and N. Lewis, "Intel Pro 1000XF Server Testing," *Penn State Behrend Electrical and Computer Engineering Senior Design Report*, 2003.

Far02   J. Farrell, Programming Logic and Design, 2nd ed., Thomson Learning, 2002.

Fis04   K. J. Fisher, M. J. Lobaugh, and D. H. Parente, "An Assessment of Desired Business Knowledge Attributes for Engineering and Technology Graduates," submitted to the *Journal of Engineering Technology*, July 2004.

Gar91   J. C. Garland, "Advice to Beginning Physics Speakers," *Physics Today*, pp. 42–45, July, 1991.

Gob99   A. Gobold, "What Has Ethics to Do with Me? I Am An Engineer," *Engineering Management Journal*, vol. 9, no. 2, pp. 81–86, 1999.

Gra85   J. Gray. "Why Do Computers Stop and What Can Be Done About It?," *Tandem Technical Report 85.7*, 46 pp., 1985.

Gra02	C. F. Gray and E. W. Larson, *Project Management*, 2nd ed., McGraw-Hill, 2002.
Gri93	A. Griffin and J. R. Hauser, "The Voice of the Customer," *Marketing Science*, vol. 12, no. 1, pp. 1–27, 1993.
Hau88	J. R. Hauser and D. Clausing, "The House of Quality," *Harvard Business Review*, pp. 63–73, May/June, 1988.
Hna03	E. R. Hnatek, Practical Reliability of Electronic Equipment and Products, Marcel Dekker, 2003.
Hym98	B. Hyman, *The Fundamentals of Engineering Design*, Prentice-Hall, 1998.
IEEE Std 610.12-1990	*IEEE Standard Glossary of Software Engineering Terminology*, IEEE Standard 610.12, 1990.
IEEE Std 830-1998	*IEEE Standard for Software Requirements Specifications*, IEEE Standard 830, 1998.
IEEE Std 1233-1998	IEEE Guide for Developing System Requirements Specifications, IEEE Standard 1233, 1998.
Jal97	P. Jalote, *An Integrated Approach to Software Engineering*, 2nd ed., Spring-Verlag, New York, 1997.
Joh02	D. W. Johnson and F. Johnson, *Joining Together: Group Theory and Group Skills*, 8th ed., Prentice Hall, 2002.
Kat93	J. R. Katzenbach and D. K. Smith, The Wisdom of Teams: Creating the High-Performance Organization, HarperCollins, 1993.
Kei84	D. Keirsey and M. Bates, *Please Understand Me: Character and Temperament Types*, Prometheus Nemesis, 1984.
Kel01	T. Kelley and J. Littman, The Art of Innovation: Lessons in Creativity from IDEO, America's Leading Design Firm, Doubleday, 2001.
Leo95	"The World-Class Engineer," Penn State University Leonhard Center, 1995.
Loc97	"The Ethics Challenge Leaders Guide," Lockheed Martin Corporation, 1997.
Mci02	"The Ten C's for Evaluating Internet Resources," University of Wisconsin-Eau Claire McIntyre Library, http://www.uwec.edu/library/Guides/tencs.html, June, 2002.
Mcl93	G. F. McClean, "Integrating Ethics and Design," *IEEE Technology and Society Magazine*, vol. 12, no. 3, pp. 19–30, 1993.
Mic91	M. Michalko, *Thinkertoys*, Ten Speed Press, 1991.
MIL-DBK 217F	Military Handbook for Reliability Prediction of Equipment, United States Department of Defense, 1991.

MIL-HDBK 881	*Department of Defense Handbook: Work Breakdown Structure,* United States Department of Defense, 1998.
Nov00	G. Novacek, "Designing for Reliability, Maintainability, and Safety: Part 1 Getting Started," *Circuit Cellar,* Issue 125, December, 2000.
Nov01	G. Novacek, "Designing for Reliability, Maintainability, and Safety: Part 2: Digging Deeper," *Circuit Cellar,* Issue 126, January, 2001.
Osb48	A. F. Osborn, *Your Creative Power: How to Use Your Imagination,* Scribner, 1948.
Osb63	A. F. Osborn, *Applied Imagination: Principles and Procedures of Creative Problem-Solving,* Scribner, 1963.
Oul04	*The Glossary of Software Vulnerability Testing,* The University of Oulu, Department of Electrical and Information Engineering webpage, http://www.ee.oulu.fi/research/ouspg/sage/glossary, 2004.
Pag88	M. Page-Jones, *The Practical Guide to Structured Systems Design,* 2nd ed., Yourdan Press Computing Series, 1988.
Pet97	T. Peters, *The Circle of Innovation: You Can't Shrink Your Way to Greatness,* Knopf, 1997.
Pet00	J. F. Peters and W. Pedrycz, *Software Engineering: An Engineering Approach,* John Wiley and Sons, 2000.
Pri90	S. Pritchard, "Friends, Romans, Cost Engineers, . . . Can We Talk," *Cost Engineering,* February, 1990.
Pug90	S. Pugh, *Total Design: Integrated Methods for Successful Product Engineering,* Addison-Wesley, 1990.
Ros93	R. Rosenburg, "The Engineering Presentation–Some Ideas on How to Approach and Present It," *IEEE Transactions on Professional Communication,* vol. PC-26, no. 1, pp. 191–193, December, 1993.
Roy70	W. W. Royce, "Managing the Development of Large Software Systems: Concepts and Techniques," *Proceedings of the IEEE WESTCON,* 1970.
Rum98	J. Rumbaugh, I. Jacobson, and G. Booch, *Unified Modeling Language Reference Manual,* Addison-Wesley, 1998.
Run99	D. L. Runnels, "How to Write Better Test Cases," International Conference on Software Testing and Review, 12 pp., 1999.
Sat88	T. Satay, *Decision Making for Leaders: The Analytical Hierarchy Process for Decisions in a Complex World,* University of Pittsburgh, 1988.
Sat02	J. W. Satzinger, *Systems Analysis and Design in a Changing World,* Course Technology, 2002.

Sch98	K. Schrock, "The ABCs of Website Evaluation," *Classroom Connect*, pp. 4–6, December 1998/January, 1999. Available at *http://kathyschrock.net/abceval/*.	
Sed04	A. Sedra and K. C. Smith, *Microelectronic Circuits*, Oxford University Press, 2004.	
Sha94	J. Shapiro, "Profession/Profile of George Heilmeier," *IEEE Spectrum*, pp. 56–58, June, 1994.	
Slo91	P. Sloane, *Lateral Thinking Puzzlers*, Sterling, 1991.	
Slo93	P. Sloane and D. MacHale, *Challenging Lateral Thinking Puzzles*, Sterling, 1993.	
Slo94	P. Sloane, *Great Lateral Thinking Puzzles*, Sterling, 1994.	
Smi04	K. A. Smith, <u>Teamwork and Project Management,</u> 2nd ed., McGraw-Hill, 2004.	
Som01	I. Sommerville, <u>Software Engineering,</u> 6th ed., Addison Wesley, 2001.	
Sta01	J. Stadtmiller, *Electronics: Project Management and Design*, Prentice Hall, 2001.	
Ste89	M. A. Stettner, <u>"How to Speak so Facts Come Alive,"</u> *Chemical Engineering*, pp. 195, September, 1989.	
Ste99	W. P. Stevens, G. J. Myers, and L. L. Constantine, "Structured Design," *IBM Systems Journal*, vol. 13, no. 2, pp. 231–256, 1999 (reprinted from 1974).	
Str02	C. E. Stroud, *A Designer's Guide to Built-In Self Test*, Kluwer Academic Publishers, 2002.	
Tel96	Telcordia SR332, *Reliability Prediction Procedure for Electronic Equipment*, 1996.	
Tuc65	B. Tuckman, "Developmental Sequence in Small Groups," *Psychological Bulletin*, vol. 63, pp. 384–399, 1965.	
Tuf03	E. Tufte, "PowerPoint Is Evil: Power Corrupts, PowerPoint Corrupts Absolutely," *Wired Magazine*, September, 2003. Available at http://www.wired.com/wired/archive/11.09/ppt2.html.	
Ulr03	K. T. Ulrich and S. D. Eppinger, *Product Design and Development*, 3rd ed., McGraw-Hill, 2003.	
Van01	A. Van Gorp and I. Van De Poel, "Ethical Considerations in the Engineering Design Process," *IEEE Technology and Society Magazine*, vol. 20, no. 3, pp. 15–22, 2001.	
Vol03	*Volere Requirements Specification Template*, Version 9, 2003.	
Wak00	J. Wakerly, *Digital Design: Principles and Practice*, 3rd ed., Prentice Hall, 2000.	
Wil95	D. Wilemon, "Cross-Functional Teamwork in Technology Based Organizations," *IEEE Engineering Management Conference*, pp. 74–79, 1995.	
Wol01	W. Wolf, *Computers and Components: Principles of Embedded Computing System Design*, Morgan Kaufmann, 2001.	
Wol02	W. Wolf, *Modern VLSI Design*, Prentice Hall, 2002.	

Yal01   S. Yalamanchili, *Introductory VHDL: From Simulation to Synthesis*, Prentice Hall, 2001.

You89   E. Yourdan, *Modern Structured Analysis*, Yourdan Press Computing Series, Prentice Hall, 1989.

Index

ABET, 4
Abler, Robert, 63
   abstract, 112
   abstractness property, 37
   acceptance test, 135, 136, 146, 156
ACM, 30
   activity diagram, 126, 131
   activity view, 126
   activity, project plan, 198, 199, 200, 201, 202, 203, 206, 209, 213
Adams, James, 66, 68, 71, 81
   adjourning, 185, 194
   analog-to-digital converter (test of), 143
Analytical Hierarchy Process, 81
Apple Newton, 18
   applied research, 19
   artifact, 18
Association for Computing Machinery, 216
   association relationship, 124
Attributes, 120
   audience, 237, 238, 240, 243
   audio graphic equalizer, 109
   audio power amplifier (design of, 92
   automated script test, 145
   availability requirement, 48

Baird, John, 243
   baseline requirements, 50
   bathtub curve, 161
Beer, David, 243
   benchmarking, 60
   bicycle computer, 110
   black box tests, 135, 137
Bohrbugs, 140
   bottom-up, 88, 89, 90, 94, 109
   brainstorming, 68, 69, 72, 80, 81, 83
   break-even analysis, 203, 204, 205, 211

buffer amplifier, 93
bug, 140, 141, 153

Cagan and Vogel, 33
Camus, Albert, 213
   cardinality ratio, 120
Challenger, 227
Chip Center, 29
Circuit Cellar, 29, 183, 231
   claims, for a patent, 218
   class diagram, 123, 131
COCOMO, 208, 213
   cohesion, 88, 106, 107, 109, 153
   collaboration diagram, 127
   combination system, 181
   communication standards, 41
Compendex, 30
   component diagram, 128
   composition relationship, 124
   concept evaluation, 75
   concept fan, 74, 75, 81, 82
   concept generation, 7, 65, 67, 69, 80, 81
   concept table, 73, 74, 81, 82
   conditional rule-based ethics, 215
   connector standards, 41
   consensus, 184, 188, 189, 192, 193, 195
   constraints, 4
Constructive Cost Model (COCOMO), 208
   controllability, 135, 138, 154, 155
   coordination module, 99
   copyrights, 213, 217, 218, 224, 225
   cost estimation, 203, 206, 213
   coupling, 88, 106, 107, 109, 111, 153
   creative designs, 18
   creativity, 65, 66, 67, 68, 69, 70, 81
   critical path, 200, 201, 202, 209, 211, 212, 213

Cross, 5, 16
Cross, Nigel, 109
Cumulative Distribution Function, 159, 161

data dictionary, 119
data flow diagrams, 111, 117, 131–132
data flow, 117
data store, 117
Database Software System Design Report, 32
database system for a college, 120
Datasheet Catalog, 29
de St-Expurey, Antoine, 91
DeBono, Edward, 65, 68, 69, 81
   deliverables, of a project plan, 198, 199, 208, 209, 211
Delphi Technique, 193
   deployment diagram, 129
   derating, 174, 175
   descriptive process, 5
Descriptive processes, 5
   design architecture, 89
   design phase, 9
   design process, 4
Design processes, 12–13, 4
   design space, 23
   detailed design, 89
   development level (of a standard), 40
DFD, 117, 118, 119, 131
DigiFridge, 124
   digital compass, in robot design, 149
Digital Refrigerator, design of, 122
   digital stopwatch design, 96
   digital stopwatch, 111
DILBERT, 24, 71, 156, 226, 236
diode, 183
Dr. Dobbs, 29

Eames, Charles, 90
  economic constraints, 43
Edison, Thomas, 111
EE Times, 29
Einstein, Albert, 235
  electromagnetic Interference, 47
Electronic Design Magazine, 29
  embedded system, 17
Embedded systems, 10
  embedded systems, 5
  emotional blocks (to creativity), 67
  energy requirement, 44
  engineer, 3
  engineering design, 4
  engineering requirement, 35, 36, 37, 38, 51, 52, 53, 62
  engineering tradeoff matrix, 59, 60
Engineering-marketing tradeoff matrix, 58
  entity relationship diagram, 111, 119, 131–132
  entity relationship matrix, 120
  environmental blocks, 67
  environmental requirements, 47
  ethical and legal constraints, 45
  ethics, 213, 214, 215, 217, 227, 228, 232, 234, 235, 236
  event space, 156
  event table, 119
  event, 119, 156
  exponential density, 160
  external event, 119
Extreme Programming, 11, 17

  failure function, 162
  failure rate, 155, 161, 162, 163, 165, 167, 168, 170, 172, 175, 176, 178, 180, 182, 183
  finite-state machine, 96, 98
  fixed costs, 203
  float, 200, 201, 202, 211, 212, 213
  flowchart, 111, 115, 116, 117, 134
  forming, 185, 186, 190, 194, 195
  functional decomposition, 87, 88, 89, 90, 91, 92, 96, 99, 100, 107, 108, 109, 111, 117, 129
  functional requirement (specification), 89
  fundamental research, 19

Gantt chart, 202, 203, 208, 211, 213
Garland, James, 243
  generalization relationship, 123
Golden Rule, 214
Gray and Larson, 211
Gray, James, 154
Griffin and Hauser, 33

Hauser and Clausing, 63
Hayes, Robert H., 16
H-bridges, in robot design, 149
  health and safety constraints, 45
  heat sink, 173, 174, 175, 184
Heilmeier, George, 20
Heisenbugs, 140, 141
  high gain amplifier, 93
  high-performance team, 186, 187
HOQ, 61, 63
House of Quality, 61
HP DeskJet printer, 205

IDEO Corporation, 24, 67
IDEO, 16, 24
IEEE Code of Ethics, 213, 216, 228, 230, 231, 232
IEEE Std. 1233–1998, 36, 37, 63
IEEE Xplore, 30, 40
IEEE, 227
  implementation level (of a standard), 40
  innovation, 3, 7, 15, 65, 66, 81, 112
  iNPD - integrated new product development, 33
  input module, 99
  integration test, 135, 146, 149, 154, 155, 156
Intel Pro 1000XF Server Testing Design Report, 31, 32
Intel Pro1000 XF, 31, 32
  intellectual and expressive (barriers to creativity), 68
  intention, 113
  interaction diagrams, *131*
  interaction view, 127
  interface, of DFD (aka source & sink), 117
  introduction (of presentation), 237, 238, 242, 243
Inverter circuit, 178

Jalote, Pankaj, 109
JANTX quality factor, 168, 170, 172
JPEG-2000 standard, 40
  judgment, 65
  junction temperature, 170, 172

Katzenbach and Smith, 184, 186, 187, 188, 194
Keirsey Temperament sorter, 187
Keirsey, 194
Kelley, Michael, 16
  key attribute, 121
KLOC (cost estimator), 207, 209

  lateral thinking, 68
  lateral thinking puzzles, 82
Law of Large Numbers, 156
LCD, in robot design, 149
Level 0, 89, 92, 94, 96, 103, 108
Level 1, 89, 93, 96, 97, 103, 104, 108, 117, 118
Level 2, 95
  liability, 217, 225, 226, 232
Librarian's Index, 28
Lockheed Martin Corporation, 228
  logic inverter, 170

  maintenance phase, 9
  majority voting system, 185
  manufacturing constraints, 46
  marketing requirements, 25
  matrix test, 142, 144, 145, 148, 151, 152, 155
MCU, in robot design, 149
  mean time to failure (MTTF), 165, 167, 170, 172, 176, 178, 179, 183
  mean, 155, 158, 159, 165, 167, 183
  mediation, 192
  meeting agenda, 189
Mehrabian, Albert, 236
  meta-standards, 41
Meyers-Briggs, 187
  microcontroller, in robot design, 149, 155
  microprocessor, 183
MIL-HDBK-217F, 40, 168, 170, 178, 182, 183, 184
MindTools, 212
  mission, 20
  module, 89
  morals, 213, 214, 232, 233

Morton-Thiokol, 227
   multi-disciplinary team, 184, 195
   mythical man-month, 208

NASA, 227
   needs statement, 30, 62
   negligence, 213, 225, 226, 230, 233
   network diagram, 199, 200, 201, 202, 208, 211, 212, 213
NGT, 72
Nominal Group Technique, 72, 80, 189
   non-disclosure agreement, 224
   normal density, 159
   norming, 185, 186, 194
Novacek, George, 183, 231

object type, 113
object, 123
objective tree, 25, 26
objectives statement, 31
object-oriented design (OOD), 123
observability, 135, 138, 149, 154, 155
of engineering design, 4
Ohgo, Norio, 87
Ohm's Law, 173
ON Semiconductor, 29
Online Ethics Center, 227
   op amp, 184
   oral presentation checklist, 242
   oral presentations, 235, 236, 243
Osborn, Alex, 69, 71, 81
   output module, 99
   output power, 53
   over-specificity, 50

pairwise, 26, 34
pairwise comparison, 26, 27, 34
PalmPilot, 18
parallel system, 179, 180, 181
patents, 213, 217, 218, 221, 222, 223, 224, 232, 233
path-complete coverage, 142
perceptual blocks (to creativity), 66
performing, 185, 186, 187, 192, 194, 195
Peters, Tom, 17

physical diagram, 131
physical view, 128
political constraints, 48
potential team, 186, 187
power dissipation, 173
power output stage, 93
predecessors, 199, 212
prescriptive design process, 5
prescriptive design, 5
principles, 214, 215
probability density function, 157
probability operator, 156
probability theory, 156
Problem Identification, 7, 17
   problem statement, 18, 24, 30, 33, 34
Problem Statement, 184, 188
   processes, of data flow diagram, 117
   processing path, 142
   product realization process, 5
Project Excellence, 15
project manager, 208, 209
   properties of the Requirements Specification, 50
   prototyping and construction phase, 9
   pseudo-team, 186, 187
Pugh Concept selection, 81
Pugh, Stuart, 63

QFD, 61, 63
Quality Functional Deployment, 61, 155

RAID, 184
   range finder, in robot design, 149
   real team, 187, 192
   redundancy, 179, 180, 182, 183, 184
Redundant Array of Independent Disks, 180
   reliability standards, 41
   reliability, 40, 48, 155, 161, 168, 170, 175, 176, 180, 182
   reliability, mathemtical definition, 161
Requirements Specification checklist, 62
Requirements Specification, 7, 17, 35, 36, 37, 49, 50, 51, 52, 53, 56, 62, 63, 64, 136, 184, 188, 241
   research phase, 7

reverse-engineering, 224
robot, 37, 38, 39, 64
robot, design of, 149, 155
Robust Design, 155
   routine designs, 18
   rross-functional team, 184
   rule-based ethics, 215
Runnel, Diane, 155
Ruskin, John, 155
Ruth, Babe, 183

safety standards, 41
Sarbanes-Oxley Act, 228
satisfice, 231
SCAMPER, 70, 81
   selection criteria, 20
   sequence diagram, 128
   series system, 176
   situational ethics, 215
Six-Sigma, 155
   slippage, 200, 208
SMART, 21
Smith, Karl, 194
   social constraints, 48
Sommerville, Ian, 16
   spiral model, 11
Spiral Software design process, 17
Stadtmiller, Joseph, 109
   standards, 36, 39, 40, 41, 45, 63, 64, 65
   state diagram, 111, 113, 114, 115, 126, 129, 131, 133, 134
State diagrams, 132
   state event, 119
   state machine view, 126
   state, 113
   static view, 123
   step-by-step test, 144
   storming, 185, 186, 194, 195
   strengths and weaknesses analysis, 76
   strict liability, 213, 225, 233
   structure charts, 99, 100, 101
   structured design, 99
   stubs, 139
   superclass, 123
   sustainability constraints, 45
   system integration, 9
   systems test, 18

team performance goals, 184
Team Process Guidelines, 183, 193, 194, 196
team process guidelines, 192
team roles, 190
team, definition of, 184
team, self-assessment, 195
Telcordia, 168
temporal event, 119
test coverage, 142
Test Vee, 136, 140
testable, 138
testing standards, 41
thermal resistance, 172, 173
thermometer design, 101
Thomas Register, 29
top-down design, 88
top-down, 88, 89, 90, 91, 109
tort, 225, 233
total harmonic distortion, 52
traceability property, 37
trade secrets, 213, 217, 224, 233
transform module, 99
transistor amplifier, test of, 138, 139
transistor, mttf, 165
transistor, reliability, 163
trigger, of an event, 119
TRIZ, 81
Tuckman, Bruce, 185
Tuft, Edward, 239

UL, 45
Ullrich and Eppinger, 63
Ulrich and Eppinger, 23, 24, 33
under-specificity, 50
Unified Modeling Language, 111, 122, 131, 132
uniform density, 159
unit test, 135, 141, 142, 143, 151, 154, 155, 156
United State Patent and Trademark Office, 217, 232
US Bureau of Labor Statistics, 30
US Government Official WebPortal, 30
US Patent Trademark Office, 218
use-case diagram, 124, 131
use-case view, 124
user level (of a stnadard), 40
utilitarian ethics, 215

validation, 51, 63, 64, 113
values, 213, 215, 217, 228, 229, 232, 233
variable costs, 203
variance, 158, 183
variant designs, 18
vending machine, 114, 115, 133
vending machine, design of, 114
verification, 37, 113
video browsing system, design of, 117
vision, 20
VLSI, 5, 9, 17
voice of the customer, 23

Wallberg, John, 234
waterfall model, 11
whistleblowers, 227, 228
white box tests, 135, 137
work breakdown structure, 212
workbreak structure, 198, 201, 202, 206, 208, 209, 211
working group, 186
World-Class Engineer, 12, 14